Mathematical Reasoning: The History and Impact of the DReaM Group

Greg Michaelson
Editor

Mathematical Reasoning: The History and Impact of the DReaM Group

Editor
Greg Michaelson
Mathematical and Computer Sciences
Heriot-Watt University
Edinburgh, UK

ISBN 978-3-030-77881-1 ISBN 978-3-030-77879-8 (eBook)
https://doi.org/10.1007/978-3-030-77879-8

This Springer imprint is published by the registered company Springer Nature Switzerland AG
The registered company address is: Gewerbestrasse 11, 6330 Cham, Switzerland

Foreword

AI, Automated Reasoning and Mathematics: DReaM

Edinburgh, one of the most beautiful cities of the world—Athens of the north—and the birthplace of Artificial Intelligence in Europe! Yes, this evokes fond memories of many pleasant visits, research exchanges, a sabbatical, and the close cooperation with Alan Bundy's research group over so many years.

Little did I know what was to come—let alone the meaning of the two letters A and I—when I came to England to study for an MSc in Computer Science at Essex University in 1972. But it was going to change my life, and all my plans, forever.

It was here that I learned for the first time that there are researchers who believe that computers can think, and that computers can have an almost human dialog about a children's blocks world. I also learned that computers can do mathematics—the queen of intellectual disciplines as I used to think as the highbrowed, newly graduated, math student I used to be—in the sense of proving novel mathematical theorems. So when Pat Hayes came from Edinburgh to become a lecturer for AI at Essex University and agreed to accept me as his PhD student, I turned down my chance to go to Oxford to be supervised by Dana Scott and stayed on: a decision I have never regretted—and it came in handy that I had a girlfriend who was doing politics at Essex in Colchester who later became my wife.

Edinburgh, Thou City Fair and High *Nisi Dominus Frustra, Psalm 127*

We learned in our AI courses that a system can do better than heuristic search, and I remember writing an essay on "What is wrong with GPS" (not the Global Positioning System, but Herb Simon's General Problem Solver based on search by a single mechanism just like a resolution theorem proving system). Pat had written his Logician's Folly paper,[1] and so, coming back to Germany after my PhD in 1976, I wanted to set up a research group for theorem proving, but not as a search-based resolution system—the still dominating paradigm at CADE. The new battle cry initiated by Carl Hewitt's thesis at MIT, which hallmarked the paradigm change in AI research, was: knowledge based systems. So, our theorem proving system was still based on resolution, but it was to be guided by a supervisor, where the mathematical knowledge should be represented. We promised our funding agent, the DFG, two things: firstly that we could build a system that was not fundamentally characterised by blind search, mathematically expressed by the R-value (the ratio of the number of clauses in the final proof divided by the total number of clauses generated in the search space); and second, that the system would be by some order of magnitude stronger than any other system on the market, as it was *knowledge based.*

After many years of development, we could show that indeed this was possible, and for many years we had a friendly race with the then strongest system on the market by Larry Wos and his research group at Argonne National Lab, Chicago. This would work by sending each other problems we had solved and hoped that the other one could not do. So sometimes Larry would call in the middle of the night, unaware of the time difference between the continents: "Hey Siekmann, can you do this?" and then we had a week or two, to show that his theorem, indeed followed from the axioms he had also sent. By and by we knew the strength and weaknesses of the Argonne system pretty well and so we sent him a problem that was a real challenge for him, but not for us, for example, as we used a sorted logic among other special features. So our message "Hey, Larry, can you do this?" went to the other side of the Atlantic—but sooner or later his smart students found the trick and solved it as well. This went on for some years with our noses still up in the air, since our theorem prover could always do as well as theirs, but with a significantly better R-value.

But, the total amount of computation including the supervisor was expensive, and in one of the panel discussions Larry provoked us with something like this: "Look, Siekmann,[2] why don't you discard your supervisor and replace it by the strongest and best system of the world, namely our system OTTER, let it find the proof and guide your base system smoothly to the proof with an R-value of exactly 1". This was good thinking, in particular since our second promise to the DFG,

[1]Bruce Anderson and Pat Hayes. The logician's folly, DCL Memo 54.University of Edinburgh. 1972.

[2]He always used my family name as opposed to the usual American custom of addressing a friend by his first name. This was to tease me with his quirky sense of humour that would take too long to explain here (for example, Larry was blind and sued Playboy Magazine for discriminating against the blind by not having any touch sensitive issues). We were really good friends, who respected each other very much.

the fundamental increase in strength, did not really come about: sometimes the pendulum swung to our side of the Atlantic but it always eventually swung back to the American side—the "knowledge base" revolution was never in sight. Also the guidance of the supervisor did not really work as we had hoped, and so by the end of the funding period we were somewhat disconcerted.

And this is the point where Alan Bundy and his ideas came to the rescue.[3]

His paper "A Science of Reasoning" impressed us deeply and left a lasting impression: could we not abandon the whole idea of the supervisor and build a system on very different principles that would come much closer to the way a human mathematician would prove a theorem? That is, to plan a proof at a more abstract level and then refine it down to the final syntactic logical proof?

Proof Planning was born!

Unusually enough, we were given a second chance by the DFG for another "Sonderforschungsbereich", which meant another 12 years of continuous funding, and this was spent on our new system OMEGA, which we considered our final word on the issue of theorem proving. Research was from now on dominated by the close collaboration and friendly competition with Alan's research group in Edinburgh: one of the most pleasant research periods in my academic life.

As a matter of principle, my PhD students had to spend at least half a year abroad, more often than not in Great Britain, and Edinburgh was their favourite spot. My late second wife, Erica Melis, even spent a whole year in beautiful Scotland with Alan Bundy's group and was full of enthusiasm not only about the wonderful countryside, but even more about the inspiring and open research atmosphere, the weekly discussion group, the blue notes, of which she wrote a few herself, and not least by Alan's advice and constructive criticism. And so, by and by, over the years to come, we had so many student exchanges that we jokingly applied to the Saarland Government to install a direct flight from Saarbrücken Airport to Edinburgh to accommodate for the exchanges. Well, there is still no direct flight and I missed many more flights and meetings in Edinburgh, because of the gruesome traffic conditions.

But it is not all work that springs to mind when I think of our visits to Edinburgh: the beautiful Firth of Forth (a tongue twisting pronunciation test for us non-natives); the walk up to Arthur's seat when you needed a break from work; relaxation during a stroll through Princes Street Gardens between the Old Town and the New Town (built in 1767 as J Strother Moore, an American working with Alan in Bernhard Meltzer's Meta-Mathematics Unit, noticed in awe); the Indian meal at Haymarket; and Erica's enthusiasm about the theatre scene during the Fringe Festival, where she took me often, sometimes more than once in a day. And the unforgettable olfactory bliss of malted barley from the city's breweries with scents of roasted malt from the North British Grain Distillery always reminding you that you are in Edinburgh. As a matter of fact, an Edinburger, as the natives are called in Auld Reekie, is no child of sadness: going out for evenings to have a beer somewhere at the Grassmarket (I must have been one of the few foreign members of CAMRA, the Campaign for

---

[3]Bundy, A.: A Science of Reasoning, pp. 178–198. MIT Press (1991).

Real Ale) or whisky tasting at the Scotch Malt Whisky Society in Queen Street—fond memories indeed. For many years we had a competition in our group, who has the smallest membership number of the Scotch Malt Whisky Society, and with now only an ϵ away from my eighties, I used to win until one day when we hired a new RA we could not believe it: he even had a significantly smaller number. This had to be celebrated with a special single cask whisky.

But it was the friendly intellectually inspiring atmosphere in Alan's research group that was the best experience: everyone was ready for a quick witticism about

Alan Bundy was appointed CBE in the 2012 New Year Honours for services to computing science

almost anything, including ourselves. It was here that I learned the best biting British humour about the royal family, but we all fell silent when we saw the following photograph:

The Beat Generation; Jack Kerouac, On the Road; Allen Ginsberg's Howl and Burrough's "Naked Lunch"; and then the1968 student demonstrations in Berlin, London and Paris and now that: Alan Bundy in his cut in front of the queen shaking her hand and receiving the CBE!

Well—even Bob Dylan received the Nobel Prize.

But it was not just automated theorem proving, as this volume shows, but many more areas in AI where Alan and his students made lasting contributions and turned Edinburgh—the Lighthill Report not withstanding—back into one of the most influential and outstanding research sites for symbolic AI in the world: reconnoitring unexplored jungle, as Mateja put it. Or to phrase it using Alan's more sober and modest words:

> *My group has been characterised by its diversity of approaches to the representation of and reasoning with knowledge, including: deduction; meta-level reasoning; learning, especially of new reasoning methods; representation creation and change; as well as applications to problems as diverse as formal verification, analogical blending and computational creativity.*

Yes, Alan you were more dominant and present in our life here at Saarbrücken than you probably know!

Congratulations on your lifetime achievements!

Saarbrücken, Germany
March 23, 2021

Jörg Siekmann

Preface

Overview

This timely and engaging collection constitutes a festschrift for the internationally leading DReaM Group (Discovery and Reasoning in Mathematics), founded and led by Alan Bundy, at the University of Edinburgh, from 1971 to the present. In many ways, of course, this book is also a festschrift for Alan himself, whose vision and leadership shine throughout.

Alan, with characteristic modesty, was adamant that the Group as a whole should be celebrated. Nonetheless, it is fundamental to acknowledge his pioneering roles, in both Mathematical Reasoning research, and in sustaining a cutting edge group driven by a strong ethos of mutually supportive inquisitiveness. Alan's contributions have been widely and deservedly recognised. Amongst other awards, he is an elected Fellow of The Royal Society (2012) and of The Royal Society of Edinburgh (1996), a Fellow of the Association for Computing Machinery (2014), and won the 2007 IJCAI Award for Research Excellence and Herbrand Award for Distinguished Contributions to Automated Deduction. In the 2012 New Year Honours, he was appointed CBE for services to Computing Science.

Under Alan's leadership, the DReaM Group enjoyed continuous funding from 1982 to 2019, from the United Kingdom Science and Engineering Research Council (SERC), and its successor, the Engineering and Physical Sciences Research Council (EPSRC). Unlike the more common time-limited, focused funding, this support was based on first Rolling, and then Platform, Grants, to underpin relevant research within a liberal interpretation of Mathematical Reasoning. This has enabled unparalleled support for exemplary activity, as is clear from the strength and diversity of the work presented here.

EPSRC ended funding for all Platform Grants in 2017, so we thought that was a good point at which to create a lasting record of the DReaM Group's contributions. However, as well as heralding the Group's substantial technical achievements, we also wished to highlight how these were nurtured by its constitution, so we asked contributors to present their achievements in wider personal and Group contexts.

Thus, where other festschrifts separate out scientific accounts from personal appreciations, we have sought to explicitly integrate the technical and the social.

In 2018, after Group discussion, I sent out a call to the DReaM email lists soliciting participation. All submissions were accepted, and the authors reviewed each other, with additional reviewers, acknowledged below. Thus, while I am formally the editor, this is very much a collective, Group endeavour.

Contents

The chapters are presented in a rough chronological order of the authors' first engagements with the DReaM Group.

In Chap. 1, Alan Bundy provides a thorough account of the Group's genesis and progress, highlighting key research themes and achievements. This chapter strongly situates the rest of the book.

In Chap. 2, J Strother Moore recounts his impressions of the environment in which the DReaM group first developed, in the Metamathematics Unit and the Department of Machine Intelligence and Perception, in the early 1970s.

In Chap. 3, Toby Walsh surveys his participation in the elaboration and formalisation of the key DReaM approach of proof planning, and of the core rewrite technique of rippling.

In Chap. 4, Paul Jackson explores systematic techniques for dynamically presenting proofs to varying degrees of detail, to aid accessibility and comprehension.

In Chap. 5, Jacques D. Fleuriot discusses the application of proof planning to nonstandard analysis, and fusing mechanical discovery and proof for effective geometric reasoning.

In Chap. 6, Gudmund Grov, Andrew Ireland, and Maria Teresa Llano describe reasoned modelling for system design, where the ideas of proof plans are extended to incorporate patterns of formal modelling.

In Chap. 7, Mateja Jamnik discusses the formalisation and automation of diagrammatic reasoning, and how best to choose representations for optimal human understanding.

Finally, in Chap. 8, Fiona McNeill explores how reasoning about failure, another central DReaM technique, can be applied to matching and integrating ontologies.

Edinburgh, UK Greg Michaelson

Acknowledgements

I wish to thank:

- Our additional reviewers Lilia Georgieva (Heriot Watt University) and Grant Olney Passmore (Imandra Inc. and University of Cambridge), for their thoughtful comments.
- Our Springer editors: Ronan Nugent, for enthusiastically embracing this project, and Paul Drougas, for patiently seeing it through to completion.
- Jörg Siekmann, for his heartfelt foreword, which ably sets the tone for the rest of the book.

Contents

Contributors

Alan Bundy School of Informatics, University of Edinburgh, Edinburgh, UK

Jacques D. Fleuriot Artificial Intelligence and Its Applications Institute (AIAI), School of Informatics, University of Edinburgh, Edinburgh, UK

Gudmund Grov Norwegian Defence Research Establishment (FFI), Kjeller, Norway

Andrew Ireland Heriot-Watt University, Edinburgh, UK

Paul B. Jackson University of Edinburgh, Edinburgh, UK

Mateja Jamnik Department of Computer Science and Technology, University of Cambridge, Cambridge, United Kingdom

Maria Teresa Llano Monash University, Melbourne, Australia

Fiona McNeill University of Edinburgh, Edinburgh, UK

Greg Michaelson Mathematical and Computer Sciences, Heriot-Watt University Edinburgh, UK

J Strother Moore The University of Texas at Austin, Computer Science Department Austin, TX, USA

Jörg Siekmann Universität des Saarlandes/DFKI Saarbrücken, Germany

Toby Walsh University of New South Wales, Sydney and Data61, Sydney, NSW, Australia

Chapter 1
The History of the DReaM Group

Alan Bundy

Abstract I describe the history of the DReaM Group (Discovery and Reasoning in Mathematics), which I created after my arrival at the University of Edinburgh in 1971. The group has been characterised by its diversity of approaches to the representation of and reasoning with knowledge, including: deduction; meta-level reasoning; learning, especially of new reasoning methods; representation creation and change; as well as applications to problems as diverse as formal verification, analogical blending and computational creativity. From 1982, we have been supported first by a series of EPSRC rolling grants and then, when this funding mechanism ceased, platform grants. Now that the latter mechanism has also ceased, we felt it was time to take stock, celebrate our achievements, assess our strengths and plan our future research. This history lays the bedrock for that self-analysis. Inevitably, space restrictions have forced me to be highly selective in what research I cover. I apologise to those whose excellent research I have had to omit or only hint at. My selection has been mainly influenced by my desire to illustrate our methodological and application diversity. I hope that the other chapters in this book will fill some of those gaps.

1.1 Why DReaM?

I have been engaged in DReaM Group research for well over four decades. I once turned down a five-fold salary increase in order to continue doing so. I have worked well past my official retirement date. Over 100 fellow researchers have enthusiastically contributed to the Group in those four decades. Why this enthusiasm?

I love mathematics, especially logic. I am also fascinated with cognition—not especially in the cognition of humans or of other animals, but how cognition is

A. Bundy (✉)
School of Informatics, University of Edinburgh, Edinburgh, UK
e-mail: A.Bundy@ed.ac.uk

G. Michaelson (eds.), *Mathematical Reasoning: The History and Impact of the DReaM Group*, https://doi.org/10.1007/978-3-030-77879-8_1

even possible. Emulating cognition using logical reasoning then ticks all my boxes. Such emulation cannot just be limited to deduction though. It is clear that cognition goes well beyond this and involves a multitude of interacting reasoning processes: abduction; analogy; planning; uncertainty; meta-level reasoning; learning; the use of diagrams; conjecture formation; fault diagnosis; representation formation and evolution, etc.

There is a long-standing argument as to whether computing is a branch of engineering or of science. Of course, it is both. It can also be mathematics. What I have been most drawn to, however, is *computing as art*. Art values beauty, of course, and many computing solutions are beautiful, including ingenious algorithms and elegant representations. For me, though, the main attraction of art is *surprise*: especially pointing to new ways of thinking. As Sir William Lawrence Bragg said:

> "The important thing in science is not so much to obtain new facts as to discover new ways of thinking about them".

Artificial intelligence is well suited to this. It often demonstrates that an aspect of cognition that most people thought was beyond automation can, in fact, be automated. The DReaM Group has had more than its fair share of such demonstrations. We have shown how: intermediate lemmas can be constructed; search control can be reasoned about; proofs can consist of diagrams; interesting conjectures can be made; concepts can be analogically blended; models of the environment can be invented and faulty ones repaired; paintings, music and poems can be created. Even more surprisingly, mathematics, often in the form of logic, can play a key role in the automation of aspects of cognition that seem inherently informal.

It has been a delight to write this chapter and remind myself of these many achievements of the DReaM Group.

1.2 Arrival in Edinburgh

I first came to Edinburgh in June 1971, having just finished a PhD in Mathematical Logic under Reuben Louis Goodstein at Leicester. I joined Bernard Meltzer's Metamathematics Unit (MMU), which was a major centre for research on automated theorem proving, and was loosely attached to the Department of Machine Intelligence and Perception (DMIP), with which it shared accommodation in Hope Park Square. Bernard had a knack for recruiting excellent people.

Bob Kowalski: had just completed the development of Selected Literal Resolution [42] with Donald Kuehner and was co-founding the field of logic programming [41].

Pat Hayes: was pioneering John McCarthy's programme of representing common sense reasoning using logic [53].

Bob Boyer & J Moore: J's original PhD project was automating the understanding of children's stories supervised by Donald Michie, but switched to automated theorem proving. BobB joined MMU at the same as me, and he and J formed the closest working partnership I have ever known, which led to the Boyer–Moore series of inductive theorem provers [6].

Anecdote 1 (The Dangers of Being Over-Rehearsed) *I first met Bernard when he came to give a seminar at Leicester. I wanted to find a job where I could use my logical knowledge but apply it to something practical. So I rehearsed a little speech, which I imagined would go like this?*

Me: "Do you have any logicians in your group?"
Bernard: "No."
Me: "That's what I thought. Are you looking for one?"

Unfortunately, how it actually went was:

Me: "Do you have any logicians in your group?".
Bernard: "No. I suppose that was obvious from my talk."
Me: "That's what I thought. Are you looking for one?"

Fortunately, Bernard immediately knew what had happened and thought it very funny.

I came to Edinburgh at a turbulent time for DMIP. It was composed of three research groups in three separate sites, the heads of which did not get on. MMU became embroiled by the ensuing spate of heated argument and serial reorganisation. This turmoil was fuelled by the Lighthill Report [48], which passed a damning verdict on AI, leading to an AI Winter, during which funding was hard to get. Many researchers left for new pastures, including BobK, Pat, BobB and J. In 1974, out of this ferment, a new, but very small, Department of Artificial Intelligence (DAI) crystallised. To make us viable, the University created two new lectureships. I was lucky to get one and Gordon Plotkin got the other. My first task was to organise a new undergraduate course, called Artificial Intelligence 2.[1]

1.3 The Mecho Project

MMU was funded by a fore-runner of the rolling grant scheme. The funding agency, which was then called SRC,[2] sent a panel to inspect us every few years.

[1] "2" because it was a second year course.

[2] Later to become SERC and then EPSRC.

Anecdote 2 (Clarity Considered Harmful) *The attitude of these panels was that they did not fully understand what Bernard's group was doing but thought it was good work that deserved support. For the meeting before I arrived, BobK and Pat had decided that they would make a big effort to get the panel to really understand the work. This succeeded, but the panel said "Ah! Now we understand and we've decided not to fund you".*

We got another chance, but the pressure was on to convince the panel that our research was worth funding.

Bernard managed to reinvigorate his research group with new people and there was talk of a group project to consolidate us. So at the next SERC panel meeting, I described such a project in which we would solve mechanics problems stated in English. The panel suggested that I submit my own grant proposal to do this. I did, and I got two tranches of SERC funding for a total of 6 years. We called our program *Mecho* (for Mechanics Oracle). We took A Level applied mathematics problems about particles, inelastic strings, friction-less pulleys, inclined planes, etc., translated the English into a first-order logic representation, then extracted and solved equations [12]. Martha Stone,[3] Chris Mellish and Rob Milne did PhDs with me to translate English into FOL. Lawrence Byrd, George Luger and I automated the extraction of simultaneous equations from the FOL, and Bob Welham and I automated solving the equations. In retrospect, this was the beginning of the DReaM Group,[4] although we did not call it that then. We were the *Mecho Group*. Martha and Chris both developed international reputations for NLP. George is best known for his successful AI textbook. Laurence joined Quintus, a Silicon Valley, logic program-based start-up founded by David Warren. BobW joined Hewlett Packard in Bristol. Rob became an expert-system entrepreneur in Scotland, but sadly died at 48 climbing Mount Everest—the last of his "Seven Summits" challenge.[5]

BobK had brought Prolog to Edinburgh and, with the work of David Warren at Edinburgh, it became a practical programming language. The Mecho Group adopted it for all the phases of the Mecho program, and it became the biggest Prolog program in the world. Members of the group, including Chris Mellish, Lawrence Byrd, Richard O'Keefe and Leon Sterling, contributed some of the key Prolog textbooks [19, 63, 71].

Meta-level reasoning was a major focus of the Mecho project. It was used to control search in natural language understanding, common sense inference, representation formation and algebraic manipulation. Meta-level reasoning had been much discussed in Bernard's group as a way to impose methodological hygiene by separating object-level reasoning from the heuristics used to control search. It was in

[3] Now Palmer.

[4] AKA—the Mathematical Reasoning Group.

[5] Climbing the highest peaks in each of seven continents.

sharp contrast to the expert systems then being built, which incorporated heuristics as additional conditions into rules, thus making it hard to separate soundness and completeness from pragmatism. In particular, proof methods, such as isolation, collection and attraction, were used to control the algebraic rewriting used to solve equations in our Prolog Equation-Solving System, PRESS, which can be seen as an early version of proof planning.

The limiting factor in the Mecho project was deciding how to represent as Physics abstractions the real-world objects described in English. For instance, was a ship to be represented as a particle on a horizontal plane, as used in relative velocity problems, or as a container floating on a fluid, as in Archimedes' Principle problems? In school mechanics problems, we found that these choices were hinted at using key words in the English text. This is unsatisfactory. The key skill in engineering modelling is to choose the best representation. Mecho was not the best vehicle for addressing this idealisation issue.

Anecdote 3 (The Importance of Dreaming) *In 1979, I was head hunted by Schlumberger to work on the application of expert systems to the exploration of oil. My wife, Josie and I were flown out to Ridgefield, Connecticut, wined and dined and offered 5 times my then salary. I turned them down.*

The head of their research lab had a great saying about research methodology. "To do research you need two things:

1. *You have to have a dream.*
2. *And you need to know what to do tomorrow".*

This was one reason for calling ourselves the DReaM Group, which stands for Discovery and Reasoning in Mathematics.

1.4 The Eco Project

As a new lecturer I had to attend a week-long course on teaching methods, which focused on practical skills. One of the exercises was to describe your research project. I chose to describe Mecho. After my talk, I was approached by another new lecturer on the course: Bob Muetzelfeldt. BobM was an ecological modeller in the Department of Forestry and Natural Resources. Ecological models represented the environment using differential equations to describe flows of energy between animals, plants, etc. Writing differential equations was not a common skill among biologists, so BobM's vision was to build a front end that allowed users to describe the ecological environment in biological terms, from which equations could then be automatically extracted. In Mecho, he saw an exemplar of his vision. So, BobM and I joined forces to construct what we decided to call the *Eco System*.

Eco proved to be a good vehicle for addressing the idealisation issue. Idealisation clues could be found, for instance, in the questions the ecologist wanted the ecological model to answer. If the question was about comparing the milk productivity of cattle of different ages, then classifying them into age classes was required. If it was, instead, about the productivity of different breeds, then classifying them by breed was required. Similarly, the environment might be divided into meadows *vs.* pasture *vs.* woods—or just represented as a grid.

We got two tranches of SERC funding for two RAs for 6 years in total [67]. One of the RAs was Mike Uschold, who became a founder of the ontology community and had a successful industrial career. Our initial choice of the second RA proved to be unsuccessful, so we had to replace him. BobM proposed a very bright undergraduate, who proved to be a brilliant choice. Dave Robertson was a key member of our group. At the end of the project, he became a lecturer in DAI, founding his own group on lightweight software engineering. He later went on to become first the Director of our Research Institute, CISA, then Head of the School of Informatics, and is currently the head of the College of Science and Engineering at the University of Edinburgh.

BobM's favoured programming language was Fortran, and Eco's differential equations were represented as difference equations in Fortran and solved numerically. Working with us, however, he became enamoured of Prolog and switched to that as his main programming language.

The Alvey Programme[6] was a UK response to the Japanese Fifth Generation Project.[7] Alvey initiated an "Architecture Study", and I was asked to lead a section of this entitled "Intelligent Front Ends" (IFE). These were user-friendly interfaces to software packages, whose aim was to make these packages accessible to non-computer scientists. Both Mecho and Eco were reinterpreted as IFEs. Another IFE developed in our group was Richard O'Keefe's ASA [62], which aimed to provide an IFE to statistics packages. The idea was for the user to describe an experiment requiring statistical analysis and ASA to suggest appropriate commands in a statistics package.

1.5 Building a Wider Community

1.5.1 Rolling Funding and Platform Grants

From 1980, our work on automated reasoning focused more on the PRESS system [70]. We were now calling ourselves the *Press Gang*. Bernard Silver's PhD was on automating the learning of equation-solving methods from example solutions. Leon Sterling was developing proof methods for inductive proof. The work was funded by

[6]https://en.wikipedia.org/wiki/Alvey. Accessed 26.8.19.

[7]https://en.wikipedia.org/wiki/Fifth_generation_computer. Accessed 26.8.19.

two separate SERC grants. SERC persuaded me to merge these two grants into one rolling funding grant. Rolling grants are reviewed every 2 or 3 years by a visiting panel and then, if successful, extended for another 4 years. Entitled "Computational Modelling of Mathematical Reasoning", this first rolling grant started in 1982 and has rolled since then, albeit becoming a platform grant in 2002. This gave our group a great deal of security and the ability to rapidly explore new directions of research.

Anecdote 4 (Virtual Money) *The rolling grants funded not just RAs, but also equipment. We had our own network of computers and a computing officer to support us. That meant we had to decide what to buy. Unfortunately, the group members were frozen by the large sums involved and were unable to make any decisions. I, therefore, proposed that we divide all prices by 100, so that a £2,000 computer appeared to be priced at only £20. That did the trick and we rapidly reached an agreement.*

The rolling and platform grants enabled our group to grow not only in diversity but also geographically. In 1982, the rolling grant had one principal investigator and one site. In 2019, our final platform grant had eleven investigators and included three sites: Edinburgh, Heriot Watt and Queen Mary London. We try to recruit the best postdocs and postgraduate students. I learnt early in my career that the best researchers will soon want to leave and start their own research groups. Rather than be disappointed at losing a key team member, I am delighted at the opportunity to establish a new collaboration. In practice, this was achieved by adding these pathfinders as new co-investigators to the grant and exploring with them the opportunities for new collaborations. As they took new research paths, this added to the diversity of our research, but the coherence of the group was maintained because many common themes continued to underpin our research.

The range of our research also grew to encompass new areas.

Rippling: In particular, we developed the *rippling* proof method for controlling the step case of inductive proofs. It used *wave-rules* that moved and removed differences between the induction conclusion and the induction hypothesis [11] (see Sect. 1.6.2).

The Productive Use of Failure: Automatic analysis of a failed proof attempt can be used to suggest missing intermediate lemmas that enable the proof attempt to continue [36] (see Sect. 1.6.4.1).

Repair of Faulty Representations: Failures of reasoning, e.g., proving false theorems or failing to prove true ones, can be automatically analysed to suggest changes to the language of a theory [46, 47, 57] (see Sect. 1.6.4.3).

Proof Planning: Our development of proof methods and meta-level reasoning crystallised into *proof planning*: specifying proof tactics so that plan formation could be used to construct a plan for a whole proof [8] (see Sect. 1.6.5.1).

Applications of Proof Planning to Formal Methods: Since inductive reasoning is required for reasoning about repetition in both software and hardware, our automation of it could be applied to software verification [37, 49, 51], hardware verification [17] and program synthesis [27, 38, 43] (see Sect. 1.6.6.1).

Applications to Cyber Security: Inductive reasoning was also applied to discover attacks on security protocols [69]. This work later led to work on reverse engineering the implementations of the RSA PKCS♯11 API standard for smart-card protocols [35]. These projects revealed serious security flaws in these protocols (see Sect. 1.6.6.2).

Tactic Learning: By detecting patterns in successful proofs, novel and useful tactics were automatically constructed [31, 68] (see Sect. 1.6.6.3).

Theory Exploration: We developed various heuristic methods for conjecturing and proving interesting theorems [15, 20, 40, 50, 58] (see Sect. 1.6.6.4).

Diagrammatic Reasoning: We pioneered the automation of reasoning by manipulating diagrams rather than logical expressions [39, 72]. This work drew on earlier work on the constructive omega rule [2] (see Sect. 1.6.6.5).

Analogical Reasoning: We have automated analogical reasoning using both ideas of Lakatos and by applying colimits in Category Theory to achieve conceptual blending [52, 66]. These have been evaluated on both mathematics and music (see Sect. 1.6.6.6).

Computational Creativity: Members of our group pioneered computational creativity research [4, 22, 24, 32], with applications to mathematics, computer games, art and music (see Sect. 1.6.6.6).

Representing Uncertainty: We have explored mechanisms for uncertain reasoning in the FRANK System [16], which not only combines deductive with statistical reasoning, but also associates error bars with numerical data and inherits them through the inference process to the final answer. We also developed the Incidence Calculus [7], a kind of probabilistic logic in which sets of weighted possible worlds were associated with formulae rather than associating numeric probabilities directly (see Sect. 1.6.6.7).

To reflect this increasing diversity, when, in 2002, the platform grant series replaced the rolling grants, we changed the title to "The Integration and Interaction of Multiple Mathematical Reasoning Processes". This title change reflected not just the broadening of our scope from deductive reasoning to planning, analogy, learning, diagrams and failure analysis, but also to the ways these processes cooperated to become more than the sum of their parts [9]. We retained our mathematical methodology by requiring rigour in the theory of these reasoning processes, but the domain of application widened to physics, multi-agents, the arts, etc.

1.5.2 Blue Book Notes and Trip Reports

Early in the history of the DReaM Group, I initiated a series of informal notes for internal communication between group members. The first one was issued in November 1976. They were initially kept in a blue folder and entitled *blue book notes*, abbreviated BBN. Each one has a footnote on the first page that says "Notes in this series are for ϵ-baked ideas, for $1 \geq \epsilon \geq 0$. Only exceptionally should they be cited or distributed outwith the Mathematical Reasoning Group". The idea was to encourage Group members to record progress made or problems encountered, however minor. They are typically a few pages long. The purpose of the footnote was to encourage people to keep a record, but with no quality threshold. Both the upper and lower bounds on the value of ϵ were intended to be taken seriously. The circulation restriction to group members is intended to encourage people to be frank and open.

I frequently find that when I record progress, the discipline required to explain this to other people helps reveal problems that had not previously occurred to me. Contrariwise, when I record problems, the same discipline helps reveal potential solutions that also had not previously occurred to me. The notes can also sometimes serve as 0th drafts of research papers. BBN 1000 has a more extended discussion of their value. This is a note I *am* prepared to distribute outwith the group. One measure of their success is that, at time of writing, we have reached BBN 1855 and our 44th year.

At the same time, I initiated a series of trip reports. Group members attending conferences, workshops or bilateral lab visits are strongly encouraged to write a report for the benefit of those group members who were unable to attend. These are best when they are issued quickly after the trip and focus on aspects of the trip that cannot be found in published proceedings, e.g., informal discussions, pointers to related work, suggestions for future work, the future of the conference/workshop. They are not expected to be polished. I have taken to writing them during the talks. I do not find this a distraction. In fact, it helps me identify the key message of the talk so I can summarise it succinctly. Then, by the wonders of the Internet, I can even publish them on the trip home. At time of writing, we have reached TR 485.

The trip reports were initially kept in a brown folder. Now, of course, we have foregone hard copy, and both series are mounted on our group webpages, but still only accessible to group members. The choice of blue and brown folders was not accidental. You have to know that I am a grand student of Wittgenstein.

Anecdote 5 (Wittgenstein's Vision) *My PhD supervisor, Reuben Louis Goodstein, was himself supervised by Ludwig Wittgenstein. Goodstein related that Wittgenstein's research group would all go to the cinema. Wittgenstein*

(continued)

Anecdote 5 (continued)
was short sighted, but too vain to wear glasses. Consequently, he insisted that they all sat in the front row, which was a bit of a struggle for those with regular sight.

1.5.3 International Collaboration

We have always been open to collaboration beyond the DReaM Group, especially with European partners. One of the first opportunities to do this was by setting up a network of fellow researchers in inductive theorem proving. Inductive proof is needed to reason about repetition, especially in recursive or iterative programs, but also about recursive data structures and parameterised hardware. Induction, however, requires the solution of especially difficult search problems, compared, for instance, to first-order deduction. For instance, even for simple inductive theorems, it may be necessary to discover and prove intermediate lemmas that do not arise as a side effect of backwards reasoning from the goal theorem. Discovering such lemmas was usually thought to require human intervention, but our productive use of failure work, outlined in Sect. 1.5.1, showed how it could be automated.

In 1992, we received 2-year EU ESPRIT funding for the Mechanising Induction (MInd) Network, which funded a series of workshops for 10 EU sites, which we held with a similar sized group of USA researchers. This network stimulated deeper collaborations with the community of ATM researchers interested in induction. We followed up especially with some of the European sites, using British Council lab-twinning money: first, in 1993, with German sites, especially Saarbücken and Darmstadt, and then in 1995 with Italian sites in Trento and Genoa. This funding led to the series of CIAO workshops,[8] which ran from 1992 to 2013. They were initially focused on inductive reasoning but gradually widened their remit to include most of the areas listed in Sect. 1.5.1. It also funded bilateral lab visits; we benefited greatly both from visitors from the Germany and Italian sites and from having DReaM Group members visiting them. These collaborations led to our involvement in the EU Calculemus Training Network, which brought together ATP researchers with Symbolic Computation ones, and spawned the Calculemus, Mathematical Knowledge Management and Intelligent Computer Mathematics conference series. I have written about this history of European ATP collaboration in more detail in [10].

[8]http://dream.inf.ed.ac.uk/events/CIAO/.

Anecdote 6 (Cornered at Speed) *I first met Fausto Giunchiglia at IJCAI-89 in Milan, where I was the Conference Chair. I had advertised a vacant RA position and Fausto was very interested in it. He tried to meet with me, but I was so busy with committee meetings, for which IJCAIs are notorious, that we could not find a time. He offered to drive me to the airport, which seem like a good solution, but I had reckoned without his outrageous driving. He drove at enormous speed with his head turned to talk to me. To stay alive, I had to keep my eyes focused on the road ahead—and to offer him the job!*

1.6 DReaM Motifs

The breadth and diversity of the research in our growing and spreading group is outlined in Sect. 1.5.1. What has, nevertheless, held us together are several common motifs. These enable us to continue to relate each branch of our research to the others.

1.6.1 Rigour and Heuristics

Many of these motifs have their origin in Meltzer's group. For instance, one of my first lessons was the importance of separating rigorous inference and heuristic search. Logical reasoning describes a search space consisting of axioms, the theorems that can be derived from them and the rules that link them together. This search space is typically too large to search exhaustively, so heuristics are used to decide which parts of it to search and in what order. Their role is to optimise the chance that a proof is found of the target conjecture. It is not possible for such heuristics to threaten the soundness of the reasoning.

This organisation facilitates proofs of the soundness of the reasoning and the completeness of the search space. It also enables a distinction between the completeness of both the search space and of its subspace that is actually searched.

When it is represented as a logical theory, knowledge is given a semantics. Using this semantics, the truth of each axiom in the theory can then be independently established. This gives an assurance that the representation is a faithful account of what it represents.

This methodology now seems to me to be obvious and apt. One can, unfortunately, find many *ad hoc* reasoning methodologies that break it, for instance, by including heuristic conditions in rules.

An exemplar of our concern for rigour is our automation of conceptual blending. "Houseboat" is an example of a conceptual blend between "house" and "boat", in which the boat is regarded as a house. "Boathouse" is another example, but this time the boat is the *occupant* of the house. Conceptual blending is a form of analogy, and it is tempting to take an *ad hoc* approach to its automation. Our approach to automation, however, is via the concept of *colimit* from Category Theory [52]. Suppose "house" and "boat" are each represented as parent logical theories, and there is a third, general, theory of which they are both instances. The general theory can be used to determine which parts of the two parent theories are aligned and which distinct, e.g., is the boat aligned with the house or its occupants? The colimit operation then merges these aligned parts to construct the conceptual blend: houseboat or boathouse.

1.6.2 Meta-Level Reasoning

We have also extended our desire for rigour to search heuristics. *Meta-level reasoning* is the encoding of such heuristic knowledge as a logical theory in its own right. The theory whose reasoning it is then guiding is called the *object theory*. The domain that the meta-theory is describing consists of the logical expressions of the object theory, object-level derivations and the common patterns of search for constructing them.

An illustration of meta-level reasoning is our work on rippling. This proof tactic is used to rewrite a goal formula so that a given formula can be used in its proof. Rippling's precondition is that the given can be embedded in the goal, i.e., that each symbol in the given can be mapped to one in the goal, so that the nesting of the given is preserved. Table 1.1 shows a very simple example of such an embedding. The result of rippling is that a subexpression of the transformed goal matches the given.

As is usual in ATP, we work backwards from the goal. The associativity equation $(x + y) + z = x + (y + z)$ can be applied to the goal, left to right, in three different ways, only one of which makes progress towards proving it. The three alternative subgoals are:

$$((c + d) + a) + b = c + (d + 42) \tag{1.1}$$

$$(c + (d + a)) + b = (c + d) + 42 \tag{1.2}$$

$$(c + d) + (a + b) = (c + d) + 42. \tag{1.3}$$

Table 1.1 A simple example that benefits from rippling

Given	Goal	Rules
$a + b = 42$	$((c + d) + a) + b = (c + d) + 42$	$(x + y) + z = x + (y + z)$
		$u = v \implies w + u = w + v$

Subgoals (1.1) and (1.2) are unwanted, as they not only make no progress, but one makes the situation worse. Subgoal (1.3) is the only one that makes progress. The next step will be to apply monotonicity rule right to left.[9] The final subgoal, $a+b = 42$, which we will call the *target*, now matches the given. In general, the given will just be a subexpression of the target, not the whole of it.

Using rippling, we can prevent the two unwanted subgoals being inferred and only allow the wanted one. The bits of the original goal and rewritten subgoals that do not correspond to the given are annotated with orange boxes, which we call *wave-fronts*. Those bits inside the wave-fronts that *do* correspond to the given are called *wave-holes*. The embedding of the given in the (sub)goals is highlighted in red. This includes both the contents of the wave-holes and all the other parts of the (sub)goals that are not in wave-fronts.

A *wave-measure* can then be calculated on these annotations, which strictly reduces on the wanted rewriting (1.3), but does not reduce on the other two: (1.1) and (1.2). The annotated (sub)goals are shown in Table 1.2. David Basin and Toby Walsh showed how formulae can be automatically annotated with wave-fronts via a process they called *difference unification* [3].

The wave-measure is calculated as follows. Note that each wave-front can be associated with a unique subexpression of the given, e.g., $\boxed{(c+d)+\ldots}^{\uparrow}$ is associated with 42. If we write the given as a tree, then it has three levels: $=$ is at the top level, $+$ and 42 are at the middle level and a and b at the bottom level. We sum up how many orange boxes occur at each of the three levels. For instance, in the original goal, one box is associated with a and one with 42. So one is at the bottom level, one in the middle level and none at the top level. We order these scores bottom to top and write these scores in a triple [1, 1, 0], which is the measure of the original goal. The scores of the three subgoals are given in Table 1.2.

Table 1.2 An example of rippling annotation

Label	(Sub)Goals	Measure
Goal	$(\boxed{(c+d)+a}^{\uparrow})+b = \boxed{(c+d)+42}^{\uparrow}$	[1,1,0]
(1.1)	$(\boxed{(c+d)+a}^{\uparrow})+b = \boxed{c+(d+42)}^{\uparrow}$	[1,1,0]
(1.2)	$\boxed{(c+(d+a))+b}^{\uparrow} = \boxed{(c+d)+42}^{\uparrow}$	[1,2,0]
(1.3)	$\boxed{(c+d)+(a+b)}^{\uparrow} = \boxed{(c+d)+42}^{\uparrow}$	[0,2,0]
Target	$a+b = 42$	[0,0,0]

[9] Recall that we are reasoning *backwards* from the goal.

Anecdote 7 (Orange is a Strange Colour) *Why were the wave-fronts orange? The research started before the advent of data projectors. We used felt tip pens on acetate sheets on overhead projectors. By trial and error, we discovered that orange was a transparent colour, so that the formulae were visible through the orange ink.*

Unfortunately, my PhD student, Pete Madden, did not get the message and used red ink instead. The red obscured the formulae, making his talk difficult to follow.

The aim is to move the wave-fronts up through the levels and, ideally, to remove them altogether, so that the given appears as an unannotated subexpression of the final subgoal.[10] The given can then be used to complete the proof. We, therefore, order the measure tuples lexicographically, and we require rippling to strictly decrease the measure. So, [1, 1, 0] is larger than [0, 2, 0] but smaller than [1, 2, 0]. We now see that subgoal (1.1) leaves the measure unchanged, subgoal (1.2) increases it and only subgoal (1.3) decreases it. Applying the monotonicity rule now removes all wave-fronts, and the given can be used to prove this final subgoal and complete the proof.

The rewrite rules can also be annotated with wave-fronts. We then call them *wave-rules*.

Wave – Rules	**LHS**	**RHS**
$(\boxed{x+\underline{y}}^{\uparrow})+z \rightarrow \boxed{x+(\underline{y+z})}^{\uparrow}$	[1,0]	[0,1]
$\boxed{w+\underline{u}}^{\uparrow} = \boxed{w+\underline{v}}^{\uparrow} \rightarrow u=v$	[2,0]	[0,0]

The **LHS** and **RHS** columns show the wave-measures of the left-hand and right-hand sides of the rules. Note that the wave-measures strictly decrease from left to right. Rewrite rules can also be automatically annotated as wave-rules so that the wave-measures strictly decrease. During application of the wave-rules, not only must the object-level expressions match, but so must the meta-level wave-annotations. In this way, rippling is completely automated and is guaranteed to terminate.

In general, rewrite rules can be annotated as wave-rules in multiple ways. For instance, $(x+y)+z = x+(y+z)$ can also be annotated as $x+(\boxed{\underline{y}+z}^{\uparrow}) \rightarrow \boxed{(\underline{x+y})+z}^{\uparrow}$. Even though this wave-rule is directed in the opposite direction from the previous annotation, rippling will not loop. This is because the wave-annotation

[10]The metaphor is of ripples on a pond. Initially, they obscure the reflection of the surrounding countryside, but as the ripples move out, the reflection is restored.

must match. So, for instance, this reverse-directed wave-rule is not applicable to the goal in Table 1.2—nor to any of the subgoals.

1.6.3 Why Prolog?

As mentioned in Sect. 1.3, Prolog was our implementation programming language of choice for the Mecho project, and for many subsequent projects until relatively recently. Partly, this can be explained by the strong Edinburgh involvement in (a) the vision of computational logic and (b) the development of Prolog as a practical programming language. There is, however, more to it than that. Prolog is a conducive vehicle for developing meta-theories.

1.6.3.1 Meta-Level Axioms for Rippling

For instance, rippling as the successive application of wave-rules to annotated goals can be recursively defined by the meta-theory axioms:

$$Subterm(given, goal) \implies Ripple(given, goal, goal)$$

$$Embed(given, goal) \wedge Wave(wrule, goal, subgoal) \wedge$$
$$Ripple(given, subgoal, target) \implies Ripple(given, goal, target),$$

where:

- $Ripple(g, g_1, g_2)$ means that subgoal g_2 is the result of applying rippling w.r.t. given g to goal g_1;
- $Subterm(given, goal)$ means that $given$ is a subterm of $goal$;
- $Embed(given, goal)$ means that $given$ is embedded in $goal$; and
- $Wave(wrule, g_1, g_2)$ means that subgoal g_2 is the result of applying wave-rule $wrule$ to goal g_1.

These axioms can be readily expressed as the Prolog program:

```
ripple(Given,Goal,Goal) :- subterm(Given,Goal), !.
ripple(Given,Goal,Target) :- embed(Given,Goal),
                             wave(Wrule,Goal,Subgoal),
                             ripple(Given,Subgoal,Target).
```

Functional languages, such as ML and Haskell, are widely used to implement tactics in ATP. We have used functional and other languages in recent work. Logic programs, however, seem especially well suited to meta-level reasoning. Compare for instance, this functional definition of $Ripple$.

$$Subterm(given, goal) \implies Ripple(given, goal) = goal$$

$$Embed(given, goal) \implies Ripple(given, goal) = Ripple(given, Wave(wrule, goal)).$$

Unfortunately, *Ripple* is not a total function. It may return several results or none. Representing it instead as a relation addresses the absence of both the existence and uniqueness properties that define functionality.

1.6.3.2 Meta-Level Axioms Attraction

Similarly, the meta-level axioms that define the isolation, collection and attraction equation-solving methods of PRESS (see Sect. 1.3) can be presented relationally. Consider, for instance, *Attraction*, which moves occurrences of the unknown closer together so that they can be collected into one occurrence and then this occurrence can be isolated on the LHS of the equation. Some example attraction rewrite rules are given in Table 1.3. When applying these rules, the meta-level condition is that u and v be instantiated to terms containing the variable, say x, which is to be attracted, but w must *not* be instantiated to a term containing x. These applications will have the effect of moving the occurrences of x closer together. The distances between u and v are defined to be the size of the smallest subexpression that contains both of them. These least subexpressions are highlighted in red in Table 1.3.

The attraction proof method can be defined by the following meta-level axioms:

$$Collect(x, goal, target) \implies Attract(x, goal, target)$$

$$AttractRule(x, arule, goal, subgoal) \wedge Attract(x, subgoal, target) \implies Attract(x, goal, target)$$

where:

- $Attract(x, g_1, g_2)$ means that subgoal g_2 is the result of attracting x in g_1;
- $AttractRule(x, arule, g_1, g_2)$ means that g_2 is the result of applying attraction rule *arule* to attract occurrences of x in g_1.
- $Collect(x, g_1, g_2)$ means that subgoal g_2 is the result of collecting x in g_1.

Table 1.3 Examples of attraction rules

Attraction Rules	LHS	RHS
$Log_w u + Log_w v \rightarrow Log_w u.v$	4	2
$w.u + w.v \rightarrow w.(u + v)$	4	2
$(w^u)^v \rightarrow w^{u.v}$	3	2
$u^{v.w} \rightarrow (u^v)^w$	3	2

We can interpret these rules as saying that we keep rewriting the goal with attraction rules until either collection can be applied (exit with success) or we can no longer find any applicable attraction rules (exit with failure).

1.6.4 The Productive Use of Failure

Failure to find a proof at the first attempt is not necessarily a reason to give up. An analysis of the cause of failure can suggest a proof repair. For instance, it might suggest trying to prove an intermediate lemma that would help move the initial proof on. It might also suggest a change of representation that would enable a wanted proof to succeed (or an unwanted one to fail).

The productive use of failure has been a recurring theme in the *DReaM* group, not just in repairing broken proofs, but also in repairing broken representations, plans and programs. In particular, we have an especial interest in the methods suggested by Lakatos to cope with counter-examples to conjectures [25, 44].

1.6.4.1 Suggesting Intermediate Lemmas

Rippling can fail if, at some point, there is no wave-rule available to move the wave-fronts up to the next level. When this happens, we know a lot about the wave-rule we would like to have. This can often be enough to identify the wave-rule we would like and to prove it as a new lemma to complete the proof. We call this *the productive use of failure*. That is, we analyse the failure to work out how to unblock the failed proof attempt.

Here is a very simple example. In Peano Arithmetic [65], addition is defined recursively by the following axioms:[11]

$$\begin{aligned} x + 0 &= x \\ x + succ(y) &= succ(x + y), \end{aligned} \tag{1.4}$$

where $succ$ is a function used to construct the natural numbers, i.e., $succ(0)$ represents 1, $succ(succ(0))$ represents 2, and so on.

Suppose we now try to prove the commutativity of $+$, i.e., $m + n = n + m$. The proof will be by structural induction. Since the goal is symmetric in m and n, it does not matter which we choose as the induction variable, say m. The step case of the induction proof is

$$m + n = n + m \implies succ(m) + n = n + succ(m).$$

[11]These axioms follow Peano's spirit but have been modified according to modern practice.

Rippling is an ideal proof tactic for the step cases of inductive proofs. The given is the induction hypothesis $m + n = n + m$, and the goal is the induction conclusion $succ(m) + n = n + succ(m)$. Annotating the goal with wave-fronts gives

$$\boxed{succ(\underline{m})}^{\uparrow} + n = n + \boxed{succ(\underline{m})}^{\uparrow} \tag{1.5}$$

and annotating (1.4), the step case of the recursive definition of +, gives

$$x + \boxed{succ(\underline{y})}^{\uparrow} \rightarrow \boxed{succ(\underline{x + y})}^{\uparrow}. \tag{1.6}$$

Unfortunately, wave-rule (1.6) applies only to the RHS of (1.5) to give

$$\boxed{succ(\underline{m})}^{\uparrow} + n = \boxed{succ(\underline{n + m})}^{\uparrow}. \tag{1.7}$$

The rippling process is now blocked as we lack a wave-rule to ripple the LHS.

We know a lot about the missing wave-rule. We want its LHS to match the LHS of (1.7), and we want it to ripple the LHS wave-front out so that it surrounds the wave-hole $m + n$. That is, it should have the shape:

$$\boxed{succ(\underline{x})}^{\uparrow} + y \rightarrow \boxed{\mathscr{F}(\underline{x + y})}^{\uparrow}, \tag{1.8}$$

where $\mathscr{F}$ stands for the, as yet, unknown contents of the RHS wave-front. We can represent $\mathscr{F}$ as a meta-level variable standing for an unknown object-level term. It remains to instantiate $\mathscr{F}$ with a ground term that will enable the remaining proof to succeed.

To complete this blocked proof, two subgoals remain: (i) prove the missing wave-rule (1.8) as a lemma and (ii) complete the rippling of the step case. Second-order unification can be used to instantiate $\mathscr{F}$ as a side effect of either of these remaining proof subgoals. In either case, $\mathscr{F}$ will be instantiated to $succ$, so that the missing wave-rule is

$$\boxed{succ(\underline{x})}^{\uparrow} + y \rightarrow \boxed{succ(\underline{x + y})}^{\uparrow},$$

and the next rippling step is

$$\boxed{succ(\underline{m + n})}^{\uparrow} = \boxed{succ(\underline{n + m})}^{\uparrow}. \tag{1.9}$$

The ripple can then be completed with the monotonicity axiom of =, considered as a wave-rule:[12]

$$\boxed{succ(\underline{x})}^{\uparrow} = \boxed{succ(\underline{y})}^{\uparrow} \rightarrow x = y, \tag{1.10}$$

which reduces the rewritten induction conclusion to the given, which is the induction hypothesis.

1.6.4.2 Lakatos Methods and Counter-Examples

In [44], Lakatos describes a rational reconstruction of the history of Euler's Theorem $V - E + F = 2$ about polyhedra, where V is the number of vertexes, E the number of edges and F the number of faces. Lakatos's aim was to describe the evolution of mathematical methodology via the evolution of techniques to deal with counter-examples to false conjectures. An initial "proof" is given due to Cauchy. Counter-examples are then discovered, and a succession of repair techniques are described. They include, for instance, *strategic withdrawal* and *counter-example barring*. In strategic withdrawal, a key concept, e.g., the definition of polyhedra, is specialised to exclude some counter-examples. In counter-example barring, the negation of a concept describing some counter-examples is made into a new precondition of the conjecture.

Simon Colton and Alison Pease's Theorem Modifier (TM) [25] repaired false conjectures using some of Lakatos's methods. TM combines: the Mace counter-example finder [56] to find counter-examples to false conjectures; Colton's HR system [21] to learn, from examples and non-examples, the new concepts needed for strategic withdrawal and counter-example barring; and the OTTER theorem prover [55] to confirm that the repaired conjecture is now a theorem. An example is given in Fig. 1.1.

1.6.4.3 Suggesting Changes of Representation

Suppose a representation of the environment is faulty. We will consider two cases:

Incompatibility: The faulty representation enables us to prove (i.e., predict) something that is false (i.e., not consistent with our observations).

Insufficiency: The faulty representation fails to prove something that we observe to be true.

These two cases are dual and can be addressed in a symmetric way. Most research into representation repair works by either deleting axioms (belief revision [33]) or

[12] If you are concerned with the direction of the rewrite arrow, recall that we are reasoning backwards, so the direction of implication is right to left.

TM was given the following faulty conjecture in Ring Theory:

$$\forall x, y.\, x^2 * y * x^2 = e$$

where e is the multiplicative identity element.

Mace found 7 supporting examples and 6 counter-examples to this conjecture. Given these two sets of examples, HR invented a concept that can be simplified to:

$$\forall z.\, z^2 = z + z$$

and used it for strategic withdrawal. Otter was then able to prove the original conjecture for just those rings with this new property.

Fig. 1.1 Correcting a faulty conjecture in ring theory

adding axioms (abduction [26]). These are sometimes the most appropriate repair, but sometimes it is preferable to change the *language* of the representation, i.e., the signature of the theory. In [47], Xue Li, Alan Smaill and I applied the *reformation algorithm* to suggest such signature changes, for instance: splitting or merging of predicates or constants; adding or removing arguments to predicates. The ABC system described here combines abduction, belief revision and reformation. For technical reasons, our implementation is currently limited to Datalog-like logical theories [18]. The axioms are Horn clauses with no functions except constants. Any variables in the head of a clause must also appear in the body.

Consider, for instance, the following Datalog theory, $\mathbb{T}$, which is adapted from [33]):

$$\begin{aligned} German(x) &\implies European(x) \\ European(x) \wedge Swan(x) &\implies White(x) \\ \implies German(Bruce) & \qquad\quad \implies Swan(Bruce). \end{aligned}$$

From $\mathbb{T}$, we can infer that $White(Bruce)$:

$$\cfrac{White(Bruce) \implies}{\cfrac{European(Bruce) \wedge Swan(Bruce) \implies}{\cfrac{German(Bruce) \wedge Swan(Bruce) \implies}{\cfrac{Swan(Bruce) \implies}{\implies} \quad \implies Swan(Bruce)} \quad \implies German(Bruce)} \quad German(x) \implies European(x)} \quad European(x) \wedge Swan(x) \implies White(x)$$

This proof is by Selected Literal Resolution (SL) [42], which, for Horn clauses, has the convenient property that each resolution step is between an axiom and a goal. This has the advantage that we can apply any repair directly to the axiom involved in either the current or an earlier SL resolution step in the current branch, so we do not need to inherit the repair back up through derived clauses to an axiom. Each axiom in the proof is given on the right-hand side of the inference step. Each proof step unifies a literal in a goal with a literal in a fact or rule head. We have highlighted these literals in brown or red.

We use Kowalski form to present clauses:

$$P_1 \wedge \ldots \wedge P_m \implies Q_1 \vee \ldots \vee Q_n.$$

For Horn clauses, $n = 0$ or $n = 1$. When $n = 0$, the Horn clause is a goal. When m is also 0, it is the empty clause. The clauses on the left-hand side of the proof are all goal clauses.

Suppose, however, that we observe that the swan $Bruce$ is not white but black. The proof above proves something that is false, i.e., $\mathbb{T}$ is incompatible. This proof is unwanted and must be broken. In [33], Gärdenfors suggests repairs to the axioms, e.g., adding an exception to one of the rules:

$$x \neq Bruce \wedge European(x) \wedge Swan(x) \implies White(x).$$

But this is a unsatisfying hack. A better repair is to split $European$ into a European type of object or a European resident. $Bruce$ is *not* a European type of swan but *is* a resident in Europe. The proof can then be broken at the red unification step. Reformation is based on the unification algorithm. In the case of an incompatibility, it can break a successful unification, and in the case of insufficiency, it can enable an unsuccessful unification.

In our example, the red unification is broken by adding an extra argument to $European$ and ensuring that this argument will be instantiated to different constants when the two literals are to be unified, thus preventing it from succeeding. The two constants are named generically $Normal$ and $Abnormal$. Reformation is a purely syntactic algorithm, and we do not currently have the semantic knowledge to call them, say, $Type$ and $Resident$.

The repaired theory $\nu(\mathbb{T})$[13] is

$$\begin{aligned} German(x, y) &\implies European(x, y) \\ European(x, Normal) \wedge Swan(x) &\implies White(x) \\ \implies German(Bruce, Abnormal) & \qquad\qquad \implies Swan(Bruce), \end{aligned}$$

where the new arguments are highlighted in red. Note that the constraints of Datalog force $German$ to also be given a new argument, because any variable in the head of a clause must also appear in the body.

[13]Pronounced "new $\mathbb{T}$".

1.6.5 The Interaction of Multiple Reasoning Processes

As outlined in Sect. 1.5.1, the DReaM Group has always had an interest in a wide range of reasoning processes: deduction, learning, abduction, analogy, statistics, diagrams, failure analysis, representation formation, change of representation, meta-level reasoning, creativity, etc. We are especially interested in how different reasoning processes interact [9]. Some of these interactions have already been discussed. For instance, Sects. 1.3 and 1.6.3.2 explain how meta-level reasoning was used to control search in natural language understanding, common sense inference, representation formation and algebraic manipulation. Sections 1.6.2 and 1.6.3 explain how meta-level reasoning can guide object-level deduction. Section 1.6.4 describes how failure analysis can suggest new lemmas and, thereby, aid deduction and how it can suggest changes of representation. Section 1.6.4.2 describes the combination of counter-example finding, concept learning and theorem proving to repair faulty conjectures.

In this section, we briefly summarise some of the other interactions we have pioneered.

1.6.5.1 Proof Planning: Abstraction of Deduction

Human mathematicians often report having a plan of a proof before tackling the detailed steps. This plan helps them guide the search for a proof. It is not bound to succeed, so may require re-planning to deal with unexpected obstacles. *Proof planning* is an attempt to automate this process [8]. Proof planning was originally implemented in Prolog, by Christian Horn and Frank van Harmelen, as the *CLAM/Oyster* system [14]. This was later upgraded to an implementation in λProlog, called λ*CLAM* [29]. The current implementation is Lucas Dixon's IsaPlanner [30], which is built on top of Isabelle.

Anecdote 8 (A Language too Far) *On paper, λProlog looked like the ideal implementation language for our proof planner. For instance, its built-in, higher-order unification was just what we needed for instantiating meta-variables when speculating missing intermediate lemmas (see Sect. 1.6.4.1). It was, however, an experimental prototype complete with bugs, but without adequate maintenance support. So, when we inevitably stumbled on bugs, there was no one to fix them. At one stage, we discovered that inclusion of some redundant λProlog code could inexplicably cause a broken λ CLAM to start working again. If it subsequently stopped working, then commenting out the redundant code would affect another temporary fix.*

Proof tactics are specified, in a meta-logic, with preconditions to describe when they are applicable and postconditions to describe their effect. For example, the specifications of the wave tactic from Sect. 1.6.3.1 are:

Preconditions:

1. The given embeds into the goal.
2. There is a wave-rule that matches the goal and produces a new subgoal.
3. Any preconditions of that wave-rule are provable.

Postconditions: These postconditions are guaranteed by the successful application of a wave-rule:

1. The given also embeds into the new subgoal.
2. The wave-measure of the subgoal is strictly smaller than that of the goal.

Such specifications enable a plan to be automatically constructed by matching the preconditions of a later tactic to the postconditions of an earlier one.

Proof plans are hierarchical, i.e., a tactic can be defined in terms of subtactics, which may include recursive calls to itself. This aids understanding, as a proof can be inspected at the top level and optionally unpacked along one or more branches of the plan. We can express this hierarchical structure as a *hiproof* [28]: a graph in which the nodes are hiproofs. An example hiproof is displayed graphically in Fig. 1.2.

The specifications of tactics facilitate recovery from failure [36]. Failure occurs if the preconditions of a plan's tactic are not satisfied. Re-planning can then take place to bridge the gap between the effects of successful tactic applications and those needed by the failed tactic or by an alternative tactic. Andrew Ireland and I showed that different patterns of precondition failure suggest different kinds of

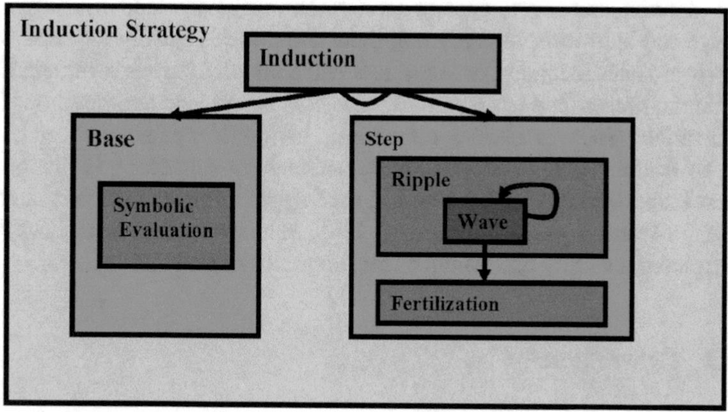

Nodes represent hiproofs; arcs represent the sub-goals passed between them.

Fig. 1.2 A hiproof of an inductive proof plan

plan repair. In inductive proofs, for instance, the analysis of failure can suggest: intermediate lemmas (see Sect. 1.6.4.1); generalisations of the conjecture, changes of the induction rule; and case analyses. Failure of precondition 2, for instance, suggests a missing intermediate lemma that could serve as the required wave-rule. On the other hand, if there is a wave-rule that partially matches, in the sense that only the wave-front fails to match, then this suggests a change of induction rule in order to ensure that the wave-front also matches.

Proof planning is just one method for abstracting proofs. In [34], Fausto Giunchiglia and Toby Walsh developed a theory of the many different forms that abstraction has taken in automated reasoning. Typically, a theory and conjecture are abstracted by ignoring some details. A proof is then found, we hope more easily, in this abstract theory. For instance, a first-order theory might be abstracted to a propositional one. This abstract proof is then used as a plan to construct a proof in the original theory.

1.6.6 Diverse Applications

The many techniques we have developed within the DReaM Group have a wide diversity of applications. They range from formal methods to cognitive science, via cyber security, mathematics, uncertainty, diagrammatic reasoning, analogy and interestingness—always combining rigour with heuristics. We highlight a few of these.

1.6.6.1 Formal Methods

Proof planning has been applied to formal verification and synthesis of both software and hardware. In particular, inductive proof is usually required whenever the software uses recursion or loops, and when the hardware is parameterised, e.g., an n-bit multiplier. One of the hardest problems in software verification is finding a loop invariant. There is a strong relationship between loop invariants and induction rules, so Andrew Ireland adapted our techniques for suggesting induction rules to suggest loop invariants [37]. Yuhui Lin used ripple failures to suggest intermediate lemmas in Event-B verification proofs [49]. As mentioned in Sect. 1.5.1, we have also explored a variety of techniques for program synthesis [27, 38, 43].

1.6.6.2 Cyber Security

Graham Steel's Coral system [69] found attacks on security protocols for group key agreement by refuting incorrect inductive conjectures. This approach avoided the need for abstraction to a group of fixed size, which can miss attacks on larger groups. By posing inductive conjectures about the trace of messages exchanged, we could investigate novel properties of the protocol, such as tolerance to disruption, and whether it results in agreement on a single key. This has allowed us to find three distinct novel attacks on groups of size two and three. This work led to the successful spin-out Cryptosense.[14]

Anecdote 9 (Just in Time) *At 1am on 8th December 2002, Graham was just finishing his PhD. Walking home from a night out when he passed our offices at 80 South Bridge and saw that they were on fire. He ran to his flat further along South Bridge and, sitting in his bay window facing onto the fire, he logged onto the servers in the basement of number 80 and downloaded his nearly complete PhD thesis. Minutes later the servers melted in the heat. The fire engulfed 11 buildings and burnt for 52 h before being completely extinguished. All the occupants of 80 South Bridge were relocated to The Appleton Tower.*

Cryptosense proposed a project to discover security flaws in smart cards [35]. Prior work had searched for flaws in the commonly used RSA PKCS♯11 API standard for smart-card protocols. In contrast, the REPROVE system automatically reverse-engineered the low-level implementations of PKCS♯11. Proprietary implementations are often used in an attempt to obtain security through obscurity. We not only showed that such implementations could be automatically decoded, but our analysis revealed extremely serious, previously unknown flaws in these implementations that severely compromised security.

A DReaM Group member, David Aspinall, is the founder of The Edinburgh Cyber Security, Privacy and Trust Institute.[15] This is a multidisciplinary research and teaching network at The University of Edinburgh.

Anecdote 10 (Men *vs.* Machines) *The Appleton Tower had won many ugliest architecture prizes. To house us, it was refurbished floor by floor—turning what had been lecture rooms into offices. We were then shuffled from floor to floor. Andrew Ireland visited me when I was on the 2nd floor and the 3rd was being refurbished. We used the lift to go to the basement, but Appleton Tower lifts have a mind of their own, so it took us to the 3rd floor instead. The doors opened on wooden boarding that prevented access to what was a building site. The doors then stuck open, which prevented the lift from moving. We were imprisoned. Fortunately, I knew exactly where to stamp to persuade the doors to close. Men 1; machines 0.*

[14]https://cryptosense.com/.

[15]https://www.ed.ac.uk/cyber-security-privacy.

1.6.6.3 Machine Learning of Proof Tactics

The proof methods used in meta-level reasoning were manually designed. This raises the question as to whether they could be learnt from examples. Bernard Silver described the application of precondition analysis successfully to learn the equation-solving methods of isolation, collection, attraction, homogenisation, etc. from example solutions [68]. The implementation was called Learning PRESS (LP). Unlike statistical machine learning techniques, precondition analysis can learn a new proof method from a single example. Like explanation-based generalisation, it generalises an example by abstracting away from the specific details of the example to construct a method that will work on similar problems. To aid it in abstraction, it has the available language of the PRESS meta-level theory.

Hazel Duncan used a statistical approach to learn Isabelle [64] tactics from proofs in Isabelle's librariesIn [31]. Proofs were abstracted into sequences of proof steps. Variable-length Markov models were then used to identify patterns that occurred more frequently than chance. These patterns were then combined by genetic programming, using a grammar of loops and splits, to form tactics. This enabled a pool of simple tactics to be formed.

1.6.6.4 When Is a Theorem Interesting?

Most applications of automated reasoning start with a conjecture and reason backwards to the axioms. If you know what theorem you want to prove, this works fine. But suppose you just want to explore a theory to discover interesting theorems? This might be a useful aid for mathematicians to explore the foothills of a new theory, while they have a coffee, to see whether it is worth taking further. Also, as an alternative to discovering new lemmas on an as-needed basis (see Sect. 1.6.4.1), you might build up a library of useful lemmas in advance.

Discovering new theorems is trivial—just reason forward from the axioms. The trick is to discover ones that are *interesting*. But what do we mean by interesting? We have conducted several experiments to address this question:

- Simon Colton compared and contrasted the measures of interestingness used in 5 machine discovery systems, extracting general principles of truth, novelty, surprisingness, non-triviality and understandability [23]. He then applied one of these 5 systems, his HR system, to the automatic invention of interesting integer sequences, seventeen of which were deemed interesting enough to have been accepted into Sloane's "Encyclopedia of Integer Sequences" [22].
- Moa Johansson's IsaCoSy [40] and Omar Montano-Rivas's IsaScheme [58] both built systems for finding interesting theorems in recursive, equational theories. They used different interestingness measures but still got surprisingly similar empirical results.

  IsaCoSy generated only equational conjectures between irreducible terms, i.e., those that a set of rewrite rules cannot simplify any further. Obvious non-theorems were rejected using a counter-example checker. Newly discovered theorems were turned into rewrite rules and added to the simplifier. This ensured that new theorems could not be proved just by simplification but required a more powerful proof method, such as induction.

IsaScheme used schemas representing common patterns of interesting theorems, such as associative, distributive and idempotency laws. New theorems were also normalised with Knuth–Bendix completion.

- MacCasland's MATHsAiD system [54] explored algebraic theories. It tried to find a balance between simplicity and non-triviality, that is, theorems were interesting iff they could be proved in a few steps but using at least one non-trivial method, e.g., simplification is trivial but induction is not. It was able to prove theorems connecting two more theories, including one about Zariski spaces that McCasland had proved in a previous career as an algebraist.

Evaluation of exploration system is challenging because there is no agreed criteria for what counts as interesting—indeed, that is what the research is aiming to investigate. The researcher's above used comparisons with previously explored theories. For instance, for IsaCoSy and IsaScheme, comparison between their outputs and Isabelle's libraries showed high values for precision and recall. For MATHsAiD, a comparison of its outputs and the theorems selected for discussion in several standard algebra textbooks showed a greater agreement between the MATHsAiD and each of the textbooks than between the textbooks themselves.

IsaCoSy is one component of the TheoryMine[16] spin-out [13]. Given an automatically generated, novel, recursive theory, it generates interesting novel theorems that customers can name.

1.6.6.5 The Constructive Omega Rule, Induction and Diagrams

The *ω-rule* is a complete, but infinitary, alternative to mathematical induction.

$$\frac{\phi(1) \wedge \phi(2) \wedge \phi(3) \wedge \ldots}{\forall x.\ \phi(x)}$$

Clearly, this is infeasible for practical theorem proving. Used backwards to prove $\forall x.\ \phi(x)$, it creates infinite branching. There is, however, a feasible version: the *constructive* ω-rule. It has the additional requirement that the $\phi(n)$ premises be proved in a *uniform* way, that is, that there exists an effective procedure, $proof_\phi$, which takes a natural number n as input and returns a proof of $\phi(n)$ as output. We will write this as $proof_\phi(n) \vdash \phi(n)$.

Cauchy's proof of Euler's formula, quoted in [44], effectively uses a polyhedral version of the *constructive* ω-rule. He describes, by example, the effective procedure

[16]theorymine.co.uk.

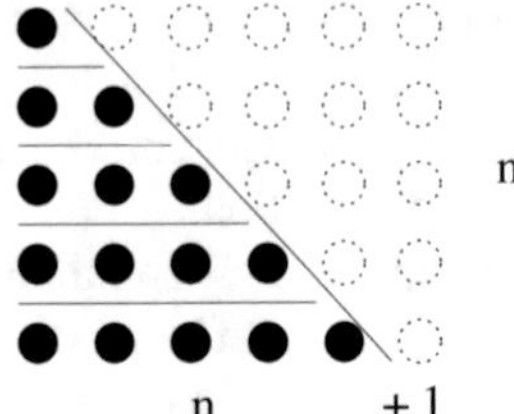

Fig. 1.3 The diagram can be viewed as a proof of the equation $1 + 2 + \ldots + n = n.(n + 1)/2$ in the case that $n = 5$. The black circles represent the LHS of the equation. The whole rectangle represents the numerator of the RHS

$proof_{V-E+F=2}$ by applying it to a cube. He then invites us to recognise its generality by claiming that it could be successfully applied to *any* polyhedron. It turns out that this claim is false—as ably demonstrated by Lakatos' many counter-examples throughout the rest of the book. Cauchy's proof omits an important step: to verify that $proof_{V-E+F=2}$ will always prove Euler's theorem for any polyhedron.

Returning to the natural number version of the constructive ω-rule, we need to prove that:

$$\forall n.\ proof_{\phi}(n) \vdash \phi(n). \tag{1.11}$$

We could, for instance, prove this by induction on n:

$$\begin{aligned} & proof_{\phi}(0) \vdash \phi(0) \\ & proof_{\phi}(n) \vdash \phi(n) \implies proof_{\phi}(n+1) \vdash \phi(n+1). \end{aligned} \tag{1.12}$$

We seem to have gone full circle: replacing induction by the constructive ω-rule only to reintroduce it in the verification of (1.11). As Siani Baker showed in [1], however, the induction required to prove (1.11) is often simpler than the induction it replaces. For instance, whereas the original induction may require an intermediate lemma, the new verification one may not.

The constructive ω-rule was also used as the basis for Mateja Jamnik's work on diagrammatic reasoning [39]. In [59], Nelsen showed how arithmetic theorems can be "proved" by displaying diagrams made of dots arranged into geometric shapes. See Fig. 1.3 for an example. Note that it exhibits only a concrete case of the theorem e.g., $\phi(5)$. The reader is expected to see that the implied special case procedure, $proof_{\phi}(5)$, can be generalised to the general one, $proof_{\phi}(n)$. Jamnik's Diamond system automated these proofs by extracting these special case proofs from the diagram, generalising them and then verifying (1.11) by induction.

1.6.6.6 Category Theory and Analogy

As outlined in Sect. 1.6.1, an *analogical blend* occurs when two old concepts are merged to form a new one. For instance, house and boat can be analogically blended in two ways: houseboat and boathouse.

A *colimit* is a construction from Category Theory that generalises constructions such as disjoint unions, direct sums, co-products, pushouts and direct limits.[17] A morphism is defined between two parent logical theories. These match entities in one parent theory to those in the other. The colimit operation is a bit like unification in that it produces a minimal super-theory of the two parent theories that respects the morphism.

The colimit construction can be implemented as a procedure that takes two parent theories and outputs this minimal super-theory. This procedure can model analogical blending. Ewen Maclean and Alan Smaill were part of the team that applied this procedure to construct new mathematical concepts from old [5]. A simple example is the construction of the integers from the natural numbers and a theory of function inversion. The same team then applied it to construct new musical concepts [32]. An example is:

> "This simple blending mechanism 'invents' a chord progression that embodies some important characteristics of the Phrygian cadence (bass downward motion by semitone to the tonic) and the perfect cadence (resolution of tritone); the blending algorithm creates a new harmonic 'concept' that was actually introduced in jazz centuries later than the original input cadences". [5, p1]

Our approach to modelling analogy is typical of our Groups' combination of Mathematical Theory and Cognitive Science.

1.6.6.7 Representing Uncertainty

The Web is a vast and rapidly growing source of information. To make best use of this information, we want, not just to retrieve known facts from it, but to combine known information from diverse sources to infer new information. To realise this ambition requires us to address several challenges. For instance, information is stored in diverse formats: databases, RDF triples, description logics and natural language. These need to be curated into a common format so that inference can combine them. Due to its size, rapid growth and constant revision, however, it is infeasible to curate it globally. We must curate it locally as we gather the information we need for each inference task. The information is also noisy and inaccurate. Inferred knowledge needs to be assigned some indication of our uncertainty in asserting it.

Query answering on the Web is an ideal vehicle for exploring the interaction of multiple reasoning processes. Deduction is not sufficient. We also want to make predictions, for instance, estimating future populations from census data. This calls for statistical reasoning techniques, such as regression or machine learning. Having formed functions by regression, we need to reason directly with them using higher-order methods, such as calculus, e.g., to estimate rates of change, points of intersection, maxima and minima, etc. We need reasoning calculi that

[17] https://en.wikipedia.org/wiki/Limit_(category_theory) accessed 15.7.19.

assign uncertainty to assertions and inherit these through the reasoning methods, including methods, such as regression, that are themselves inherently uncertain. We need representational change to curate information into a common format. We need to diagnose and repair faults in the information we retrieve. Kwabena Nuamah implemented some of these techniques in the FRANK system (Functional Reasoner for Acquiring New Knowledge) [16, 60, 61].

An earlier attempt to merge deduction and probabilistic reasoning was my *Incidence Calculus* [7]. This was proposed, as an alternative to probabilistic logics, in order to solve the problem that numeric probabilities cannot be inherited through a derivation unless they are conditionally independent. Instead, sets of weighted possible worlds were associated with formulae rather than numeric probabilities. The degree of dependence between probabilities can then be represented by the amount of overlap between these sets.

1.7 Conclusion

For more than four decades, the DReaM Group has conducted a diverse range of innovative research projects, characterised by the motifs of: rigour combined with heuristics; meta-level reasoning, the productive use of failure; the interaction of diverse reasoning techniques; the learning of new reasoning methods; representational creation and evolution; and applications of rigorous mathematics to cognitive science problems, such as analogy, music, diagrammatic reasoning, etc.

These four decades have seen a radical change in system building. In the 1970s, AI researchers built systems from scratch, typically in Lisp, Prolog or a similar declarative language. Mecho [12], for instance, was built in this way. The result was often brittle systems that were slow to develop. Nowadays, large parts of systems are constructed using third-party packages from libraries, GitHub, etc. Since these packages have often been thoroughly tested in prior applications, they tend to be robust. Development time is also faster, since you can build on the shoulders of giants rather than from first principles. Our FRANK system has been built like this [16].

There is a danger, however, as illustrated in Anecdote 8. If you rely on a package that is not well maintained, then bugs may not be fixed, and your system can become unusable. You may not discover this until you have invested a lot of energy and time into a failing system for which a replacement package is not available. So, although package use is definitely the way to go, it will repay you to carry out due diligence on each package before you commit to it. And, of course, there is not a package for every functionality you need, so you will still need to roll up your sleeves to write code to fill in the gaps.

It is instructive to contrast the kind of research programme described in J Moore's chapter in this volume, with the kind I have described above. His career has been devoted to constructing and applying a sequence of closely related theorem provers, culminating in the ACL2 prover. This can be applied to verify huge industrial-strength systems, such as microprocessors. As such, it is used routinely by several multinational hardware developers. In contrast, we have automated a diverse range of cognitive tasks building proof-of-concept prototypes that were not intended to be generic packages and were not widely used outside our group. We have been pioneers reconnoitring unexplored jungle. The six-lane highways come later. Or, you may describe our approach as "AI as art": surprising people that it is even possible to automate aspects of cognition that seemed inherently to require human interaction.

Chapter 2 contrasts the 3–4-year UK PhDs with the 5–7-year US ones. The longer PhD provides the time for robust package building, and many US PhDs follow that pattern, with the developers offering maintenance support and community building to facilitate the continued use of their package. This is "AI as Engineering". Even so, few packages attain widespread uptake, especially when the packages' developers move on from their PhDs, perhaps to work for someone with a different agenda. Thus, our group's approach to research may have been inevitable.

AI has become almost synonymous with statistical machine learning (SML). In contrast, our group's research has been mainly symbolic: the epitome of "Good, Old-Fashioned AI" (GOFAI). If we cannot beat them, should we join them: mug up on statistics and obtain some massive data? I think not. SML has some fundamental limitations:

- It *requires* huge amounts of data. Humans, in contrast, can often learn from a single example combined with background knowledge.
- It is typically used to classify objects: cats *vs.* dogs; good move *vs.* bad; mortgage or not. Sometimes, however, it is necessary to learn compound structures, e.g., computer programs, that cannot be viewed as classification tasks.
- SML systems work only with the features provided by the user. Sometimes, it is necessary to learn new concepts.
- It is inherently hard to explain the workings of a SML system. There is no logical derivation, only a huge and complex statistical calculation. But some applications *require* explanations—it may even be a legal requirement.

SML applications are beginning to confront these limitations. So, there is a future for us in developing hybrid approaches, where symbolic and statistical techniques complement each other. Launchbury describes a "Third Wave of AI" that advocates this hybrid approach [45]. Members of our group have already shown considerable enthusiasm for this third wave. Several of us attended a recent workshop at Imperial College organised by the Human-Like Computing Network+ (HLC), which has also embraced the third wave. The HLC Network+ has also created a community of AI researchers and cognitive scientists to learn from each other, which will be a crucial ingredient in the third wave, since humans can often do what is beyond SML and vice versa.

Of the work described here, that on learning new tactics from example proofs come closest to realising that third-wave vision (see Sect. 1.6.6.3). The work of my PhD students Bernard Silver [68] and Simon Colton [21] were inherently logic-based learning techniques, but Hazel Duncan's [31] used the statistical technique of variable-length Markov models. The time might be ripe to re-investigate this approach.

We are also combining deductive and statistical reasoning in the FRANK system (see Sect. 1.6.6.7). So far, our use of statistics has been confined to regression, but we have plans to explore a variety of SML techniques. Currently, FRANK explains its reasoning by abstracting its deductive reasoning and associating English text augmented with graphical displays of regression. It will be interesting to explore the potential for explanations when more complex forms of SML are interleaved with deduction.

Acknowledgments The research reported in this paper was mainly supported by a succession of EPSRC rolling and platform grants. Thanks to Greg Michaelson, Kwabena Nuamah, Predrag Janičić and two anonymous referees for feedback on earlier versions.

References

1. Baker, S.: A new application for explanation-based generalisation within automated reasoning. In: A. Bundy (ed.) 12th International Conference on Automated Deduction, Lecture Notes in Artificial Intelligence, Vol. 814, pp. 177–191. Springer-Verlag, Nancy, France (1994)
2. Baker, S., Ireland, A., Smaill, A.: On the use of the constructive omega rule within automated deduction. In: A. Voronkov (ed.) International Conference on Logic Programming and Automated Reasoning — LPAR 92, St. Petersburg, Lecture Notes in Artificial Intelligence No. 624, pp. 214–225. Springer-Verlag (1992)
3. Basin, D.A., Walsh, T.: Difference unification. In: R. Bajcsy (ed.) Proc. 13th Intern. Joint Conference on Artificial Intelligence (IJCAI '93), vol. 1, pp. 116–122. Morgan Kaufmann, San Mateo, CA (1993). Also available as Technical Report MPI-I-92-247, Max-Planck-Institut für Informatik
4. Besold, T.R., Schorlemmer, M., Smaill, A. (eds.): Computational Creativity Research: Towards Creative Machines, *Atlantis Thinking Machines*, vol. 7. Atlantis Press (2015)
5. Bou, F., Schorlemmer, M., Corneli, J., Gomez-Ramirez, D., Maclean, E., Smaill, A., Pease, A.: The role of blending in mathematical invention. In: Proceedings of the sixth international conference of computational creativity (2015)
6. Boyer, R.S., Moore, J.S.: Proving theorems about LISP functions. In: N. Nilsson (ed.) Proceedings of the Third IJCAI, pp. 486–493. International Joint Conference on Artificial Intelligence (1973). Also available from Edinburgh as DCL memo No. 60
7. Bundy, A.: Incidence calculus: A mechanism for probabilistic reasoning. Journal of Automated Reasoning **1**(3), 263–284 (1985)
8. Bundy, A.: A Science of Reasoning, pp. 178–198. MIT Press (1991)
9. Bundy, A.: Cooperating reasoning processes: more than just the sum of their parts. In: M. Veloso (ed.) Proceedings of IJCAI 2007, pp. 2–11. IJCAI Inc (2007). Acceptance speech for Research Excellence Award.
10. Bundy, A.: European collaboration on automated reasoning. AI Communications **27**(1), 25–35 (2013). https://doi.org/10.3233/AIC-130584.
11. Bundy, A., Basin, D., Hutter, D., Ireland, A.: Rippling: Meta-level Guidance for Mathematical Reasoning, *Cambridge Tracts in Theoretical Computer Science*, vol. 56. Cambridge University Press (2005)
12. Bundy, A., Byrd, L., Luger, G., Mellish, C., Milne, R., Palmer, M.: Solving mechanics problems using meta-level inference. In: B.G. Buchanan (ed.) Proceedings of IJCAI-79, pp. 1017–1027. International Joint Conference on Artificial Intelligence (1979)

13. Bundy, A., Cavallo, F., Dixon, L., Johansson, M., McCasland, R.L.: The theory behind TheoryMine. In: Automatheo. FLoC (2010)
14. Bundy, A., van Harmelen, F., Horn, C., Smaill., A.: The Oyster-Clam system. 10th International Conference on Automated Deduction (1990)
15. Bundy, A., McCasland, R., Smith, P.: Mathsaid: Automated mathematical theory exploration. Applied Intelligence **47**(3), 585–606 (2017). https://doi.org/10.1007/s10489-017-0954-8
16. Bundy, A., Nuamah, K., Lucas, C.: Automated reasoning in the age of the internet. In: 13th International Conference on Artificial Intelligence and Symbolic Computation, vol. LNAI 11110, pp. 3–18. Springer, Cham (2018). https://doi.org/10.1007/978-3-319-99957-9_1. Invited Talk
17. Cantu, F., Bundy, A., Smaill, A., Basin, D.: Experiments in automating hardware verification using inductive proof planning. In: M. Srivas, A. Camilleri (eds.) Proceedings of the Formal Methods for Computer-Aided Design Conference, no. 1166 in Lecture Notes in Computer Science, pp. 94–108. Springer-Verlag (1996)
18. Ceri, S., Gottlob, G., Tanca, L.: Logic Programming and Databases. Surveys in Computer Science. Springer-Verlag, Berlin (1990)
19. Clocksin, W.F., Mellish, C.S.: Programming in Prolog. Springer Verlag (1984)
20. Colton, S.: Automated conjecture making in number theory using HR, Otter and Maple. Journal of Symbolic Computation **39**, 593–615 (2005)
21. Colton, S., Bundy, A., Walsh, T.: HR: Automatic concept formation in pure mathematics. In: Proceedings of the 16th International Joint Conference on Artificial Intelligence, Stockholm, Sweden, pp. 786–791 (1999)
22. Colton, S., Bundy, A., Walsh, T.: Automatic invention of integer sequences. In: Proceedings of the 17th National Conference on Artificial Intelligence, Austin, Texas, USA, pp. 558–563 (2000)
23. Colton, S., Bundy, A., Walsh, T.: On the notion of interestingness in automated mathematical discovery. International Journal of Human Computer Studies **53(3)**, 351–375 (2000)
24. Colton, S., Gow, J., Torres, P., Cairns, P.: Experiments in objet trouve browsing. In: Proceedings of the 1st International Conference on Computational Creativity (2010)
25. Colton, S., Pease, A.: The TM system for repairing non-theorems. In: W. Ahrendt, P. Baumgartner, H. de Nivelle, S. Ranise, C. Tinelli (eds.) Selected papers from the IJCAR'04 disproving workshop, *Electronic Notes in Theoretical Computer Science*, vol. 125 (3), pp. 87–101 (2004)
26. Cox, P.T., Pietrzykowski, T.: Causes for events: Their computation and applications. In: J. Siekmann (ed.) Lecture Notes in Computer Science: Proceedings of the 8th International Conference on Automated Deduction, pp. 608–621. Springer-Verlag (1986)
27. Cresswell, S., Smaill, A., Richardson, J.D.C.: Deductive synthesis of recursive plans in linear logic. In: Proceedings of the 5th European Conference on Planning, Durham, UK, *LNAI*, vol. 1809. Springer Verlag (1999)
28. Denney, E., Power, J., Tourlas, K.: Hiproofs: A hierarchical notion of proof tree. Electronic Notes in Theoretical Computer Science **155**, 341–359 (2006)
29. Dennis, L.A., Jamnik, M., Pollet, M.: On the comparison of proof planning systems: Lambda-Clam, Omega and IsaPlanner. In: Proceedings of 12th Symposium on the Integration of Symbolic Computation and Mechanized Reasoning (Calculemus 2005), Electronic Notes in Computer Science (ENTCS) (2005). Available from http://www.cs.nott.ac.uk/~lad/work/publications.html
30. Dixon, L., Fleuriot, J.D.: IsaPlanner: A prototype proof planner in Isabelle. In: Proceedings of CADE'03, *LNCS*, vol. 2741, pp. 279–283 (2003)
31. Duncan, H., Bundy, A., Levine, J., Storkey, A., Pollet, M.: The use of data-mining for the automatic formation of tactics. In: Workshop on Computer-Supported Mathematical Theory Development. IJCAR-04 (2004)
32. Eppe, M., Confalonieri, R., Maclean, E., Kaliakatsos, M., Cambouropoulos, E., Codescu, M., Schorlemmer, M., Kühnberger, K.: Computational invention of cadences and chord progressions by conceptual chord-blending. In: Proceedings of the 24th International Joint Conference on Artificial Intelligence (2015)

33. Gärdenfors, P.: Belief Revision. No. 29 in Cambridge Tracts in Theoretical Computer Science. Cambridge University Press (1992)
34. Giunchiglia, F., Walsh, T.: A theory of abstraction. Artificial Intelligence **56**(2–3), 323–390 (1992)
35. Gkaniatsou, A., McNeill, F., Bundy, A., Steel, G., Focardi, R., Bozzato, C.: Getting to know your card: Reverse-engineering the smart-card application protocol data unit. In: ACSAC 2015 Proceedings of the 31st Annual Computer Security Applications Conference, pp. 441–450. ACM (2015). https://doi.org/10.1145/2818000.2818020
36. Ireland, A., Bundy, A.: Productive use of failure in inductive proof. Journal of Automated Reasoning **16**(1–2), 79–111 (1996)
37. Ireland, A., Ellis, B.J., Cook, A., Chapman, R., Barnes, J.: An integrated approach to high integrity software verification. Journal of Automated Reasoning: Special Issue on Empirically Successful Automated Reasoning **36**(4), 379–410 (2006)
38. Ireland, A., Stark, J.: Combining proof plans with partial order planning for imperative program synthesis. Journal of Automated Software Engineering **13**(1), 65–105 (2005)
39. Jamnik, M., Bundy, A., Green, I.: On automating diagrammatic proofs of arithmetic arguments. Journal of Logic, Language and Information **8**(3), 297–321 (1999)
40. Johansson, M., Dixon, L., Bundy, A.: Conjecture synthesis for inductive theories. Journal of Automated Reasoning **47**, 251–289 (2011)
41. Kowalski, R.: Logic for Problem Solving. Artificial Intelligence Series. North Holland (1979)
42. Kowalski, R.A., Kuehner, D.: Linear resolution with selection function. Artificial Intelligence **2**, 227–60 (1971)
43. Kraan, I., Basin, D., Bundy, A.: Middle-out reasoning for logic program synthesis. In: Proc. 10th Intern. Conference on Logic Programming (ICLP '93) (Budapest, Hungary), pp. 441–455. MIT Press, Cambridge, MA (1993)
44. Lakatos, I.: Proofs and Refutations: The Logic of Mathematical Discovery. Cambridge University Press (1976)
45. Launchbury, J.: A DARPA perspective on artificial intelligence (2018). Talk with slides on YouTube
46. Lehmann, J., Chan, M., Bundy, A.: A higher-order approach to ontology evolution in Physics. Journal on Data Semantics pp. 1–25 (2013). https://doi.org/10.1007/s13740-012-0016-7
47. Li, X., Bundy, A., Smaill, A.: ABC repair system for Datalog-like theories. In: 10th International Joint Conference on Knowledge Discovery, Knowledge Engineering and Knowledge Management, vol. 2, pp. 335–342. SCITEPRESS, Seville, Spain (2018). https://doi.org/10.5220/0006959703350342
48. Lighthill, S.J.: Artificial Intelligence: A General Survey, pp. 1–21. Science Research Council (1973)
49. Lin, Y., Bundy, A., Grov, G., Maclean, E.: Automating Event-B invariant proofs by rippling and proof patching. Formal Aspects of Computing pp. 1–35 (2019). https://doi.org/10.1007/s00165-018-00476-7
50. Llano, M.T., Ireland, A., Pease, A.: Discovery of invariants through automated theory formation. Formal Aspects Computing **26**(2), 203–249 (2014)
51. Maclean, E., Ireland, A., Grov, G.: Proof automation for functional correctness in separation logic. Journal of Logic and Computation (2014). https://doi.org/10.1093/logcom/exu032
52. Martinez, M., Abdel-Fattah, A., Krumnack, U., Gomez-Ramirez, D., Smaill, A., Besold, T.R., Pease, A., Schmidt, M., Guhe, M., Kühnberger, K.U.: Theory blending: Extended algorithmic aspects and examples. Annals of Mathematics and Artificial Intelligence **80**(1), 65–89 (2017). https://doi.org/10.1007/s10472-016-9505-y. URL http://homepages.inf.ed.ac.uk/smaill/martinezEtAl16.pdf
53. McCarthy, J., Hayes, P.: Some philosophical problems from the standpoint of artificial intelligence. In: B. Meltzer, D. Michie (eds.) Machine Intelligence 4. Edinburgh University Press (1969)
54. McCasland, R.L., Bundy, A., Smith, P.F.: Ascertaining mathematical theorems. In: J. Carette, W.M. Farmer (eds.) Proceedings of Calculemus 2005. Newcastle, UK (2005)

55. McCune, W.: The Otter user's guide. Tech. Rep. ANL/90/9, Argonne National Laboratory (1990)
56. McCune, W.: A Davis-Putnam program and its application to finite first-order model search. Tech. Rep. ANL/MCS-TM-194, Argonne National Laboratories (1994)
57. McNeill, F., Bundy, A.: Dynamic, automatic, first-order ontology repair by diagnosis of failed plan execution. International Journal on Semantic Web and Information Systems **3**(3), 1–35 (2007). Special issue on ontology matching.
58. Montano-Rivas, O., McCasland, R., Dixon, L., Bundy, A.: Scheme-based theorem discovery and concept invention. Expert Systems with Applications **39**(2), 1637–1646 (2012)
59. Nelsen, R.B.: Proofs without Words: Exercises in Visual Thinking. The Mathematical Association of America (1993)
60. Nuamah, K.: Functional inferences over heterogeneous data. Ph.D. thesis, University of Edinburgh (2018)
61. Nuamah, K., Bundy, A.: Calculating error bars on inferences from web data. In: SAI Intelligent Systems Conference (IntelliSys), pp. 618–640. Springer, Cham (2018). https://doi.org/10.1007/978-3-030-01057-7_48
62. O'Keefe, R.: Logic and lattices for a statistics advisor. Ph.D. thesis, University of Edinburgh (1987)
63. O'Keefe, R.A.: The Craft of Prolog. MIT Press, Cambridge, Mass (1990)
64. Paulson, L.: Isabelle: the next 700 theorem provers. In: P. Odifreddi (ed.) Logic and Computer Science, pp. 77–90. Academic Press (1990)
65. Peano, G.: Arithmetices Principia Novo Methodo Exposita. Bocca (1889). URL http://eudml.org/doc/203509
66. Pease, A., Colton, S., Ramezani, R., Smaill, A., Guhe, M.: Using analogical representations for mathematical concept formation. In: L. Magnani, W. Carnielli, C. Pizzi (eds.) Model-based Reasoning in Science and Technology: Abduction, Logic, and Computational Discovery, no. 341 in Studies in Computational Intelligence, pp. 301–314. Springer (2010). URL http://springerlink.com/content/y1t348758g46q462/fulltext.pdf
67. Robertson, D., Bundy, A., Muetzelfeldt, R., Haggith, M., Uschold, M.: Eco-Logic: Logic-Based Approaches to Ecological Modelling. MIT Press (1991)
68. Silver, B.: Meta-level inference: Representing and Learning Control Information in Artificial Intelligence. North Holland (1985). Revised version of the author's PhD thesis, Department of Artificial Intelligence, U. of Edinburgh, 1984
69. Steel, G., Bundy, A.: Attacking group protocols by refuting incorrect inductive conjectures. Journal of Automated Reasoning **First Online Edition**, 1–28 (2005). Special Issue on Automated Reasoning for Security Protocol Analysis
70. Sterling, L., Bundy, A., Byrd, L., O'Keefe, R., Silver, B.: Solving symbolic equations with PRESS. J. Symbolic Computation **7**, 71–84 (1982). Also available from Edinburgh as DAI Research Paper 171.
71. Sterling, L., Shapiro, E.: The Art of Prolog. MIT Press, Cambridge, MA (1986)
72. Winterstein, D., Bundy, A., Gurr, C.: Dr. Doodle: A diagrammatic theorem prover. In: D. Basin, M. Rusinowitch (eds.) Automated Reasoning, Lecture Notes in Computer Science, pp. 331–335. Springer Berlin Heidelberg (2004). https://doi.org/10.1007/978-3-540-25984-8_24

Chapter 2
Recollections of Hope Park Square, 1970–1973

J Strother Moore

Abstract I reminisce about the time and place the DReaM Group started: the Metamathematics Unit and its sister, the Department of Machine Intelligence and Perception, in Hope Park Square, Meadow Lane, Edinburgh, in the early 1970s. This is not meant as a scholarly history but just a personal recollection that tries to capture the spirit of the times.

2.1 Arrival

"2 Hope Park Square, Meadow Lane, Edinburgh" read the address on my letter of acceptance. I could not imagine a more romantic address, nor a more improbable one for an academic department with the equally improbable name "Department of Machine Intelligence and Perception." It was September, 1970, and I had just arrived from MIT where I had gotten my bachelor's degree in mathematics. Now I was going to start my PhD studies. My academic career seemed to keep taking me back in time. In Texas, where I grew up, a building was old if it was built in the 1930s. Cambridge, Massachusetts, where MIT is located, pushed "old" back to the eighteenth century. But then there was Edinburgh, whose "New Town" was started in 1766. Parts of Hope Park Square predated that.

Hope Park Square was a collection of stone buildings surrounding an overgrown garden that you entered through an archway. The buildings housed both Machine Intelligence and Perception and the Metamathematics Unit.[1] I went through the arch and entered the building to be greeted by the Servitor—what a strange idea to have a door man for an academic department. "I'm here to see Rod Burstall," I

[1]The Metamathematics Unit, whose name was soon changed to the Department of Computational Logic, became a founding member of the School of Artificial Intelligence.

J. S. Moore (✉)
The University of Texas at Austin, Computer Science Department, Austin, TX, USA
e-mail: moore@cs.utexas.edu

G. Michaelson (eds.), *Mathematical Reasoning: The History and Impact of the DReaM Group*, https://doi.org/10.1007/978-3-030-77879-8_2

explained, and was directed up the stairs. Rod had signed my letter of acceptance, and so it seemed natural to start with him. The next person I met turned out to be his secretary, Eleanor Kerse, who pointed me to his door. I knocked and upon being invited in I found him standing on his head. "I'll be done in a few minutes," he said.

After introductions, Rod took me to a nearby office with two desks. "There are pencils and paper in that cupboard. This sheet explains how to log on to the timesharing system." He also handed me a slim silver book, "This describes the programming language, POP-2." He did not note that he was a co-author of it, along with Robin Popplestone.[2] POP-2 was an elegant Lisp-like programming language with an Algol-like syntax. "We have tea every day at 11 am and 4 pm. Welcome to Hope Park Square." And then he left.

My research career had begun. My first surprise was how different the British educational system was from the American one I had experienced. My office mate in the Machine Intelligence department was Mike Gordon, who like me was a new PhD student that year. From Mike, I learned that my undergraduate education at MIT differed greatly from his at Cambridge. We both majored in Mathematics, but at MIT I was required to take a broad range of science and technical courses and consequently less maths, whereas at Cambridge Mike focused almost entirely on maths. But the biggest difference in the systems, as I came to experience it, was in the PhD programs. At Edinburgh, there were no classes, no lectures, no homework, no exams—except the oral exam for my dissertation. And, at least in my case, no direction. While I was not aware of it at the time, a PhD in Edinburgh was typically completed in 3 years, as opposed to 5–7 in the USA. I have come to appreciate this shortened time span because it can enable more imaginative dissertation topics! You are unlikely to solve the problem in 3 years so you can aim high and try to demonstrate the plausibility of an approach. While there were no formal classes, Hope Park Square hosted a steady stream of outstanding scientists who gave seminars on cutting edge research in AI and theorem proving—Marvin Minksy, John McCarthy, Christopher Strachey, Dana Scott, J. Alan Robinson, Woody Bledsoe, Danny Bobrow, Robin Milner. In addition, tea time was invariably interesting and educational, with Burstall, Popplestone, Donald Michie, Harry Barrow, our visitors, and half a dozen PhD students including Mike, Gordon Plotkin, and John Darlington.

My plan in 1970 was to build a natural language understanding system that could read and answer questions about children's stories. My adviser was Donald Michie.

Donald and I spoke often in those early days. He sometimes talked about his days at Bletchley Park, even though in 1970 that work was still classified. I now know he worked on the German "Tunny" encryption, which was the British code name for the Lorenz cipher, which was considered harder than Enigma. The work

[2] I got to know Robin soon after that. He had a sailboat berthed in Leith and wanted it in Inverness so he could take his family down the Great Glen, through Loch Ness to Fort William. But he did not want to sail his family through the North Sea. He needed someone to crew for him and I volunteered. I had spent a lot of time in boats growing up—mainly canoes on the bayou—but occasionally in dinghies and sail boats in Galveston Bay.

at Bletchley Park was not declassified until after I left Edinburgh, so Donald never told me much. But he did say they did not have the code machine and so they built a simulator for it. And I asked a really naive question without thinking: "Did you do that in software or actually build a physical machine?" He had to remind me: they were inventing digital computers then. "Software" was a thing of the future.

Meanwhile, I needed a job in Edinburgh. I had financed most of my MIT education as a computer programmer. I learned to program, in Fortran, in 1964, while in high school, by taking a Summer Science Training course sponsored by the National Science Foundation. My first job at MIT was in the MIT Laser Research Group, where I was hired to fetch and carry glass tubes, electrodes, machined glass, etc., from the physics stores and machine shops to the lab. I was not very good at that—I would often bring back the wrong thing. And while I was waiting for my next assignment, I would watch the grad students try to figure out what was wrong with their Fortran programs. "You need a comma there." Or, "I think you mean X here." After a few weeks of unsuccessful fetch-and-carry, one of the students said "I'll go get the tubing. You try to fix this program." By my second semester at MIT, I was their programmer and I spent my time on numerical methods for solving differential equations.

But in 1968, back home in Texas for the summer, I was an intern working for TRW Systems Group at the Manned Spacecraft Center, debugging the Lunar Orbit Insertion procedure for Apollo by simulating missions: preparing card decks that put the simulator in some initial state and then simulating, second by second, 30 min of flight through variations of the events allowed by the flight plans. That summer changed my appreciation of computing: it was not just for solving differential equations. Inside a computer, a person could build a whole solar system—or any world at all. I returned to MIT, finished my math requirements perfunctorily, and focused on artificial intelligence. The next summer I helped write a page fault simulator for IBM Cambridge to explore memory management on the then-new System 360. And in my last year of MIT, I was a full-time employee of State Street Bank and Trust, Boston, working on their mutual fund software in PL/1.

So when Donald spoke of simulating the coding machine, digital simulation sprang naturally to mind. And when I needed a job in Edinburgh, the natural thing to do was to look for a programming job. I found one next door, at 9 Hope Park Square, in Bernard Meltzer's Metamathematics Unit. As far as I can recall, the Metamathematics Unit then consisted of Bernard, his quite remarkable secretary, Jean Pollock, two post-doc research fellows, Bob Kowalski and Pat Hayes, and Don Kuehner who was finishing up a PhD. We called the institution simply "the Unit," and I wore a Dr. Who pin reading "U.N.I.T." on my blue jean jacket the entire time I was in Edinburgh.

Kowalski and Kuehner had invented "Selected Literal (SL-) resolution," a restriction of resolution that was complete and enjoyed certain properties related to proof search efficiency and the shortness of proofs. Neither Kowalski nor Kuehner enjoyed programming and so I was hired to be Kowalski's programmer and my job was to create the first implementation of SL-resolution.

The job required getting a work permit. I remember sitting across from Bernard in his office and him talking on the phone to some official who would pursue the work permit. Bernard would listen to the person and then cover the phone and ask me "How long have you been programming?" "6 years," I replied. So Bernard said, "We need someone with 5 or 6 years of programming experience." Then, covering the phone, he asked me "In what languages?" "Fortran, assembly language, PL/1, and Lisp." And so it went. The eventual job description probably described exactly one person in the UK. So I got my work permit.

Bob Kowalski then described what he wanted, and I started implementing my first theorem prover. Before the first year was out, I realized I was a lot better at writing theorem provers than I was at getting computers to understand children's stories. So I went to Donald and asked if I could change advisers. He agreed and Rod Burstall, the Machine Intelligence faculty member most familiar with logic and mechanized proof, became my official dissertation adviser. I moved from the Machine Intelligence office that Rod had first shown me to an office in the Unit.

The programming environment was fairly primitive, at least by the standards of MIT in the late 1960s. Researchers at Hope Park Square shared an ICL 4130 processor with 64K bytes of 2 ms memory, three 4MB disc drives, two paper tape readers and punches, and a 300 line-per-minute line printer. It could support 8 interactive jobs, each controlled by teletype, all permanently core resident. Time sharing was strictly round-robin with all users having equal priority. Job control and programming was in POP-2, so from the user's perspective it was a "POP-2 machine" akin to but pre-dating the "Lisp machine." The CPU could be scheduled for exclusive use at night. Memory limitations drove almost all programming decisions. Teletypes connected Hope Park Square to the CPU, but all other hardware were located about 10 min' walk away at Forrest Hill.

Files were edited with a software-implemented paper tape editor, a machine that copies a paper tape by punching corresponding holes in a blank tape but permitting the user to skip sections or manually punch different holes. To edit a file you copied the bytes in it to another file, stopping at the first place you wanted to change. You would type the new text and/or delete the old text and then continue copying. You could not look at more than one byte at a time, back up, or undo. Repeated passes through the evolving file were typically required. Editing a program was tedious and error-prone. If you printed your file on the line printer to see the final result of editing, you needed to walk along the Meadows to Forrest Hill to get your output.

Memory constraints made my SL-resolution prover quite limited, and my attempts to save space by representing clauses in a compressed string format slowed it down. But by September, 1971, I could demonstrate SL-resolution on simple sets of clauses. Kowalski could ignore the performance issues but was disappointed by the size of the search space. One of his responses to this was to develop an amazing knack of creating clause sets whose refutation was straightforward, almost like the SL-resolution engine was interpreting code.

When Don Kuehner graduated with his PhD, I bought his automobile: a diesel-powered, decommissioned, Austin FX4 London taxi. It cost me £75 in 1971, and I sold it for £5 in December, 1973, because it could no longer pass its MOT inspection. In the mean time, it was an excellent vehicle for exploring Scotland—its tight turning radius made it ideal for Highland roads—and its roomy passenger compartment was great for hauling furniture. Don and I both regarded the taxi as a shared resource at Hope Park Square.

In September, 1971, two very influential people joined the Unit as post-docs: Alan Bundy and Bob Boyer.

Because Boyer and I were both from Texas, we shared an office: it was the warmest room in Hope Park Square. Our office overlooked the bowling green and practice range of the Royal Company of Archers, the Queen's ceremonial bodyguard in Scotland. At other times, there were bowlers on the green. The police bagpipe band also practiced there.

Boyer and I began to collaborate on a variety of projects including an efficient, structure-shared way to represent clauses in a resolution prover [BM72]. Instead of copying clauses to rename variables apart and implement substitution and resolvent formation, resolvents were represented by pointing to the two parents, noting which literals were resolved upon, and storing the unifying substitution. Linear proofs produced just stacks of frames, reminiscent to the way conventional programming languages implement procedure calls.[3]

In December, 1971, J. A. Robinson—who invented resolution and who made several long visits to Edinburgh—playfully awarded us one of my prized honors, a handwritten certificate that reads:

> Foundation for the Advancement of Computational Logic
> by the Taking Out of Fingers
>
> 1971 PROGRAMMING PRIZE
>
> In 1971, by the gift of an anonymous (he couldn't even remember his own name!) donor, an annual programming prize was set up, to be awarded to:
>
> "...that person, or those persons, who, in the opinion of the Board of Trustees, shall have, in the given year, contributed the most valuable, beautiful, or just plain deep and satisfying, idea to the world in the area of actual writing of programs in computational logic, as opposed to simply waving hands and hoping things will work out all right on the night...";

[3]The connection of structure sharing to Prolog and the Warren Abstract Machine is mentioned later.

> In 1971, the prize is award, by unanimous agreement of the Board, to
>
> Robert S. Boyer
> and
> J Strother Moore
>
> for their idea, explained in "The Sharing of Structure in Resolution Programs", of representing clauses as their own genesis. The Board declared, on making the announcement of the award, that this idea is "... bloody marvelous".
>
> Because of the state of the economy, and what with one thing and another, the prize this first year is somewhat less than it will be in future years.
>
> J. A. Robinson
> Secretary
> December 12, 1971

Once Boyer and I got the SL-resolution prover converted into a structured-shared clause representation, we began experimenting with proofs. But because we were passionate about programming, we tended to focus on proofs about programs.

We formalized an assembly-like language called Baroque in predicate calculus and used the SL-resolution prover as an interpreter for it. Then, programming in Baroque, we axiomatized a simple subset of Pure Lisp and could use SL-resolution to run Lisp programs and to prove simple theorems about Lisp. See pages 73–75 of my dissertation [Moo73], where, for example, I define

```
MEMBER: (MEMBER X Y) -> U
        WHERE
        (COND Y
              (COND (EQUAL X (CAR Y))
                    T
                    (MEMBER X (CDR Y)))
              NIL) -> U;
        END;
```

(At the time, we used `COND` as a 3-place if-then-else.) If one executed the statement

```
(MEMBER 2 (CONS 1 (CONS 2 (CONS 3 NIL)))) -> U;
```

the SL-resolution "interpreter" would bind `U` to `T`. More interesting to us at the time was that we could "run" programs "backwards" finding input that satisfied certain output constraints, e.g.,

```
(MEMBER 2 (CONS 1 (CONS X (CONS 3 NIL)))) -> T;
```

would bind `X` to `2`, proving "there exists an `X` such that `2` is a `MEMBER` of `(CONS 1 (CONS X (CONS 3 NIL)))`.

In Part I of my 1973 dissertation, I explored various capabilities allowed by structure sharing and "programming in the predicate calculus," including precomputing unifiers, "shallow binding" to short-circuit the dive through the stack for a binding, how to attach pragmatic restrictions on variables to prevent any instantiation from forming certain "useless" terms, and Baroque.

There were many other ramifications of structure sharing we could have explored, but our abiding interest was in proving more interesting theorems like the associativity of list concatenation, or that something is a `MEMBER` of the concatenation of two lists iff it is a `MEMBER` of one or the other. These theorems were out of reach of our SL-prover: they require induction. So we set aside resolution and structure sharing and focused on proving inductive theorems about Lisp functions.

Following McCarthy [McC63], we adopted a simple subset of Pure Lisp as our logic and took turns defining functions and proving theorems at the blackboard, questioning each move. "Why did you expand that function?" "Why induct on `A`?" "Is the conjecture general enough for induction?" As strange as it may seem today, these were groundbreaking questions in the early 1970s because "theorem proving" was almost synonymous with uniform proof procedures for first-order predicate calculus. After several months, we had a collection of heuristics to help decide when to induct and what inductive argument to use, as well as heuristics for controlling the application of axioms as rewrite rules, including the unfolding of recursively defined functions and rudimentary generalization.

We then set about implementing this in POP-2, creating the Edinburgh Pure Lisp Theorem Prover (PLTP), which was running by March, 1973, in the style of Woody Bledsoe [BBH72]. The techniques developed for PLTP have been widely adopted in virtually all modern theorem provers aimed at hardware and software verification. Indeed, I think we founded the now-thriving field of mechanized inductive theorem proving.

PLTP was a fully automatic, heuristic theorem prover for Pure Lisp, focused on induction and recursion. The user presented it with Lisp function definitions and conjectures to prove. It printed a narrative description of its evolving proof attempt. Unlike resolution provers, PLTP did not backtrack or do much search. As I said in my dissertation ([Moo73] page 208), "The program is designed to make the right decision the first time, and then pursue one goal with power and perseverance."

Its proof techniques included simplification via rewriting with Lisp axioms and definitions, heuristic use of equality, generalization, and induction. These techniques were tightly coupled so that induction set up simplification, simplification was used to determine appropriate inductions, and equality substitution and generalization were used to produce conjectures intended for inductive proofs and tended to "discover" interesting subgoals. For example, the associativity of Peano multiplication required three inductions and discovered the distributivity of multiplication over addition and the associativity of addition—a curious level of competence demonstrated too often by PLTP to be random or coincidental and explained years later by Alan Bundy's work on rippling.

Boyer and I kept a file containing all the definitions and theorems the prover succeeded on. We called it "the proveall." Every time we would change the heuristics, we ran the proveall because we had learned from past experience that it was easy to "improve" a theorem prover so that it could find a previously undiscoverable proof without us realizing it could no longer discover some "old" proofs. The proveall grew as we refined the heuristics through 1973. We established the discipline of never accepting an "improvement" until the new system had passed the proveall test. (We sometimes found that failures had more to do with peculiar aspects of the statements of "old" theorems than with faulty heuristics and restated those theorems to take advantage of new techniques.) We follow this discipline today and highly recommend it to developers of theorem provers.

Sometime during late 1971 or early 1972, we simply could not stand to use the paper tape editor to maintain our software. Inspired by our structure sharing work, we invented a way to edit a file without using much memory: build a data structure that described the edited document in terms of segments of the original file (on disk) and the text entered by the user during the edit session. We named the editor the "77-Editor" because it resided on disk track 77. We had help from D.J.M. Davies dealing with some system programming issues. The editor supported the illusion that the entire document was in memory, you could search backwards and forwards, move around in it, undo changes, etc. [BDM73]. We used the 77-Editor extensively to create PLTP.

For lunch, there was a little shop at the end of Meadow Lane that sold seven kinds of sandwiches: beans, cheese, fried egg, beans and cheese, beans and fried egg, cheese and fried eggs, and beans, cheese, and fried eggs.

When Boyer and I were stuck, we would often take walks, usually to Holyrood Park, and often the solution would come to us there. I sometimes went jogging with Alan Robinson around Holyrood Park. Once going up the road around Arthur's Seat, I said to him "I wish I had a low gear, like my bike," and he replied "You do: take smaller steps."

Some afternoons Boyer and I would go out to the Meadows behind Hope Park Square and join a pickup game of (British) football, often with school boys. They could dribble circles around the clumsy Americans. But we paid them back when we would play (American) football or baseball, both of which require throwing and catching.

Once returning from the Meadows, we found a man trying to steal Boyer's bicycle, which was parked in the archway of Hope Park Square. Boyer chased him with the baseball bat, and it is fortunate for the thief as well as for theorem proving that Boyer did not catch him. As someone said when we got back to the Unit, "A liberal is someone who's never been robbed."

In the summer of 1973, Rod Burstall came into the office and told me "You should write this up." My 3 years were ending! In the 23 months that Boyer and I worked together in Edinburgh—creating structure sharing, a text editor, and an inductive theorem prover—I am not sure that either of us thought about my PhD. We were just doing research, chasing our shared dream of automated reasoning about programs.

I wrote my dissertation that summer. Part I was about structure sharing. Part II was about PLTP. I wrote "two" dissertations because it was impossible to separate my contributions from Boyer's. The PLTP proveall then contained 47 function definitions and 67 theorems about recursive list processing (e.g., concatenation, reverse, member, union, intersection, sorting), Peano arithmetic (e.g., addition, multiplication, exponentiation), tree processing (copy, flatten, searching), and connections between them (e.g., the length of a concatenation is the sum of the lengths). All the PLTP theorems were proved completely automatically. Because PLTP was the first mechanical prover designed around induction and most of these theorems required induction, most of these theorems had never been proved mechanically before.

My oral exam was in the Fall, 1973, and my internal examiners were Bernard and Rod and my external examiners were David Cooper and Robin Milner. Milner was visiting from Stanford where he was developing LCF [Mil79]; he had read my dissertation very carefully and was especially interested in induction. We sat in Bernard's office and had tea over a teletype. They would challenge PLTP and I would explain either why it failed or how it succeeded. Some of the challenge theorems were familiar because Boyer and I had seen it prove them, e.g., that insertion sort produced ordered output. But some were proved for the first time during the oral exam, including that insertion sort preserved the number of occurrences of an element. The committee objected to my use of the non-word "normalation" as the name of the proof technique finally called "simplification." And, with a few edits to the dissertation, I was done.[4]

We left Edinburgh in December, 1973, in part because of the Lighthill Report and the coming "AI winter" in the UK. But the ideas we developed in the Metamathematics Unit lived on and had real impact. Structure sharing, which had helped reify the idea of programming in the predicate calculus, played a role in the creation and implementations of Prolog and the Warren Abstract Machine. As for our text editor, it remained in use in the Edinburgh AI community for several years, until the ICL 4130 was replaced by a PDP-10. More importantly, our document representation became an integral part of Microsoft Word.

When I left Edinburgh I joined Xerox Palo Alto Research Center (PARC). There I learned that Charles Simonyi was implementing the first WYSIWYG text editor, Bravo, for the Xerox Alto personal computer and was facing severe memory limitations. I explained our document representation to him and implemented a package of text editing utilities in Interlisp as an experimental prototype of the basic operations of search, insertion, deletion, etc. He adopted the representation in his implementation of Bravo. I maintained the Interlisp package and added features at his request even after I left PARC in 1976 to join Boyer at SRI [Moo81]. The representation not only maintains a small memory footprint and facilitates undoing

[4]PLTP was reproduced several times in the 1970s. There are at least two modern re-implementations, one by me in ACL2 and one by Grant Passmore in ML. See the PLTP Archive at http://www.cs.utexas.edu/users/moore/best-ideas/pltp/index.html.

but enables metadata, like font and change tracking, to be attached to text without changing the text. When Simonyi left PARC and joined Microsoft, he created Microsoft Word inspired in part by Bravo. He used our document representation there too, and it is still in use in Word today.

And the Pure Lisp Theorem Prover remained the focus of Boyer's and my work. In subsequent years, we explored a plethora of topics to improve the prover, including use of previously proved lemmas, verified metafunctions, integrated decision procedures, the adoption of a subset of an ANSI standard programming language as the logic, programming the system in its own logical language, and the dual use of formal models as specifications and efficiently executable prototypes. By 1979, PLTP had become Thm, the prover described in our 1979 book *A Computational Logic* [BM79]; by the mid-1980s, Thm had become Nqthm [BM88, BM97]; and by the early 1990s, Nqthm had become ACL2, *A Computational Logic for Applicative Common Lisp* [KMM00b, KMM00a, KM19]. For a sketch of that evolutionary sequence and the changes we made, see [Moo19]. I am still working on ACL2, but for the past 26 years my co-author has been Matt Kaufmann. We release a new version of ACL2 about twice a year. ACL2 is in routine use in industry including ARM, AMD, Centaur, Kestrel, Intel, Oracle, and Rockwell Collins—nightly in some cases—to verify microprocessor components and critical algorithms [WAHKMS17], dealing with conjectures that Boyer and I could not have imagined in 1973. The "proveall" has grown from PLTP's 67 theorems to ACL2's 153,823.[5]

To understand how remarkable the Metamathematics Unit had become with the arrival of Boyer and Bundy, consider this: Kowalski and Kuehner were primarily pursuing uniform proof procedures, specifically resolution in first-order predicate calculus. Early applications focused on formalizing a robot's world and using theorem proving to plan a sequence of actions to achieve some goal, following the formalization of McCarthy and Hayes' situational calculus [MH69]. Kowalski was particularly adept at formalizing clausal problems in a way that made SL proofs easy to find and that probably led to his view that one could program in the predicate calculus. Into this strictly first-order, resolution group, Meltzer added two post-docs, Bundy and Boyer, who came at mechanized reasoning from completely different perspectives. Bundy did his dissertation on proofs of Gödel's incompleteness theorems in restricted formal systems of arithmetic [Bun71], supervised by Reuben Goodstein, a master of constructive mathematics and the foundations of logic. Arithmetic is inherently inductive, and so Bundy's view of theorem proving was necessarily broader than resolution. Boyer was very familiar with resolution. His dissertation was on a restriction of resolution [Boy71]), but his supervisor, Woody Bledsoe, was a fierce advocate of *non*-uniform, heuristic provers. Boyer had co-authored a heuristic prover with Bledsoe and W.H. Henneman [BBH72] on proofs

[5]This is a conservative estimate as of July, 2019, of the number of theorems explicitly stated by users in ACL2 Community Books repository, https://github.com/acl2/acl2, which Kaufmann and I re-run for every new version of ACL2. It is conservative because it only counts conjectures presented with the `defthm` command and not conjectures required to admit definitions or presented or generated by macros, etc.

of limit theorems. And then there was me, a programmer learning theorem proving. Within a few months of Boyer's arrival at Hope Park Square, we were exploring inductive proofs about recursive data structures. Meltzer's genius is indicated by the group he assembled. We represented a wide variety of theorem proving backgrounds, styles, and applications. He basically just turned us loose.

The result was an intense, exciting, and fascinating time full of discovery. We discussed and debated everything from simple arithmetic challenges requiring generalization and induction to whether logic was an appropriate way to model human thought.

I remember discovering that some truly simple pragmas attached to "action variables" could make many of the robot problems easy, e.g., disallow the immediate composition of the action `LET-GO` onto the action `PICK-UP` or otherwise the prover would pursue the possibilities allowed by picking up an object and immediately letting it go, *ad infinitum*. Such pragmas were easily implemented in the structure-shared representation, c.f. [Moo73], page 48.

When debating whether predicate calculus could capture English, Boyer clarified the question.

Boyer: "Give me a sentence."

Kowalski: "The girl guides fish."

Boyer (writing on the black board):

At(0,'T) ∧ At(1,'h) ∧ At(2,'e) ∧ At(3,Space) ∧ At(4,'g) . . ..

On another occasion, somebody asserted "People think in predicate calculus." Somebody else said, "People think in English," to which somebody else replied, "Actually, most people think in Chinese."

Once after a tedious argument on a question I have long forgotten Kowalski said "Let's vote. And then we'll argue about who won."

Of course, mostly we discussed theorem proving ideas: hyperresolution, paramodulation, uniform versus non-uniform proof procedures, the role of soundness and completeness, how to deal with the equality relation, the role of induction, the use of lemmas, heuristics for limiting the search space, etc.

Perhaps the best picture of those years at Hope Park Square was drawn in 1971 or 1972, by Martin Pollock, FRS, husband of Bernard's secretary Jean. Martin was a founding father of molecular biology as a distinct field. He was also well known at the university for his satirical cartoons. Jean frequently had to type manuscripts that were incomprehensible to her. Below was Martin's response to a paper by Bernard.

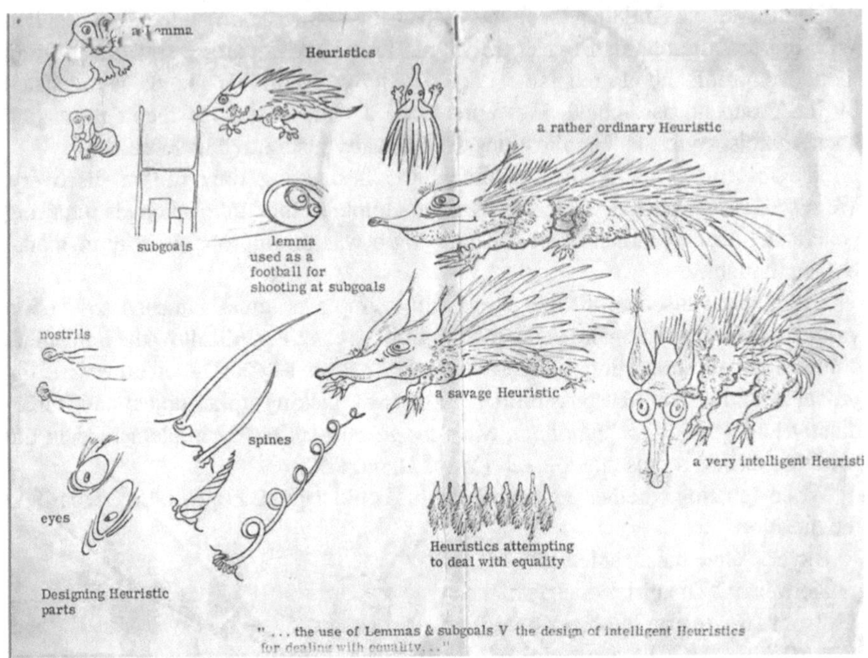

Since this book is dedicated to the research conducted by Alan Bundy's DReaM Group, it seems only fitting for me to comment on that work. Alan's chapter on the history of the DReaM group makes an interesting contrast to my history of the evolution of PLTP to ACL2 [Moo19]. There could hardly be two more different research styles: my focus was quite narrow and Alan's was extraordinarily broad. And yet it is interesting that we visited many of the same topics from our different perspectives, e.g., rippling, lemma discovery, the value of a formal meta-theory, and, of course, the concern over soundness and the impact of an *ad hoc* programming style versus a disciplined partition between heuristic and rules of inference. And in almost every case, he and I made different decisions: mine always driven by pragmatic desire to build a sound and effective prover for computational problems and his to understand how that is done. In fact, our almost half-century of pursuing-related goals from different perspectives exemplifies the wonderful atmosphere of the Unit.

References

BBH72 W.W. Bledsoe, R.S. Boyer, and W.H. Henneman. Computer proofs of limit theorems. *Artificial Intelligence*, 3:27–60, 1972.

BDM73 R. S. Boyer, D. J. M. Davies, and J S. Moore. The 77-editor. Technical Report 62, Department of Computational Logic, University of Edinburgh, 1973.

BM72 R. S. Boyer and J S. Moore. The sharing of structure in theorem-proving programs. In *Machine Intelligence 7*, pages 101–116. Edinburgh University Press, 1972.

BM79 R. S. Boyer and J S. Moore. *A Computational Logic*. Academic Press, New York, 1979.

BM88 R. S. Boyer and J S. Moore. *A Computational Logic Handbook*. Academic Press, New York, 1988.

BM97 R. S. Boyer and J S. Moore. *A Computational Logic Handbook, Second Edition*. Academic Press, New York, 1997.

Boy71 Robert S. Boyer. *Locking: A Restriction of Resolution*. Department of Mathematics, University of Texas at Austin, 1971.

Bun71 A. Bundy. *The Metatheory of the Elementary Equation Calculus*. PhD thesis, University of Leicester, August 1971.

KM19 M. Kaufmann and J S. Moore. The ACL2 home page. In *http://www.cs.utexas.edu/users/moore/acl2/*. Dept. of Computer Sciences, University of Texas at Austin, 2019.

KMM00a M. Kaufmann, P. Manolios, and J S. Moore, editors. *Computer-Aided Reasoning: ACL2 Case Studies*. Kluwer Academic Press, Boston, MA, 2000.

KMM00b M. Kaufmann, P. Manolios, and J S. Moore. *Computer-Aided Reasoning: An Approach*. Kluwer Academic Press, Boston, MA, 2000.

McC63 J. McCarthy. A basis for a mathematical theory of computation. In *Computer Programming and Formal Systems*. North-Holland Publishing Company, Amsterdam, The Netherlands, 1963.

MH69 J. McCarthy and P. Hayes. Some philosophical problems from the standpoint of artificial intelligence. In *Machine Intelligence*, volume 4, pages 463–502. Edinburgh University Press, 1969.

Mil79 Robin Milner. Lcf: A way of doing proofs with a machine. In Jiří Bečvář, editor, *Mathematical Foundations of Computer Science 1979*, pages 146–159. Lecture Notes in Computer Science, Vol. 74, Springer, Berlin Heidelberg, 1979.

Moo73 J S. Moore. Computational logic: Structure sharing and proof of program properties. Ph.D. dissertation, University of Edinburgh, 1973. See http://www.era.lib.ed.ac.uk/handle/1842/2245.

Moo81 J S. Moore. Text editing primitives – the TXDT package. Technical Report CSL-81-2 (see http://www.cs.utexas.edu/users/moore/publications/txdt-package.pdf), Xerox PARC, 1981.

Moo19 J Strother Moore. Milestones from the Pure Lisp theorem prover to acl2. *Formal Aspects of Computing*, 2019.

WAHKMS17 Jr. W. A. Hunt, M. Kaufmann, J S. Moore, and A. Slobodova. Industrial hardware and software verification with ACL2. In *Verified Trustworthy Software Systems*, volume 375. The Royal Society, 2017. (Article Number 20150399).

Chapter 3
Adventures in Mathematical Reasoning

Toby Walsh

Abstract *"Mathematics is not a careful march down a well-cleared highway, but a journey into a strange wilderness, where the explorers often get lost. Rigour should be a signal to the historian that the maps have been made, and the real explorers have gone elsewhere."*

W.S. Anglin, the Mathematical Intelligencer, 4 (4), 1982.

3.1 Introduction

In 1986,[1] I moved to Edinburgh to start a Masters conversion course into Artificial Intelligence after having studied mathematics at the University of Cambridge. I had dreamed about working in AI for many years. So it was my good fortune to fall into the gravitational attraction of Alan Bundy and become a member of the DReaM group. Shortly after, I began a PhD under Alan's careful supervision[2] [1–3].

I would now start to dream about getting computers to do mathematics. It was perhaps not surprising that this was the orbit into which I fell. I had always liked mathematics, and now I could combine two of my passions: Artificial Intelligence and mathematics.

I would stay in Edinburgh for most of the next dozen or so years, apart from some enjoyable excursions to work at INRIA in Nancy and with Fausto Giunchiglia at IRST in Trento. There was a lot to like about living and working in Auld Reekie. However, fresh challenges started to emerge, and I began to build up to an escape velocity. I took a research position in Glasgow but stayed living in Edinburgh,

[1] How can it be that long ago?

[2] I was lucky also to have one of Alan's postdocs, Fausto Giunchiglia as a second supervisor. We would work together closely for the next decade.

T. Walsh (✉)
University of New South Wales, Sydney and Data61, Sydney, NSW, Australia
e-mail: tw@cse.unsw.edu.au

G. Michaelson (eds.), *Mathematical Reasoning: The History and Impact of the DReaM Group*, https://doi.org/10.1007/978-3-030-77879-8_3

visiting the DReaM group frequently. Then I moved to York, and finally a sling shot sent me past Cork to Sydney, Australia where I remain today.

In due course, I would leave behind the work that I had done in Edinburgh and explore other parts of Artificial Intelligence such as constraint programming [4] and satisfiability [5]. However, this volume offers me the chance to consider how those ripples might have had a small influence on what followed. More importantly, it lets me thank Alan for his mentoring. This summary is necessarily high level and will avoid going into many of the technical details. I have a lot of ground to cover, so it is impossible in this limited space to go deeper.

3.2 Rippling

Alan was (and is) a neat AI researcher. One strategy he promoted was to take some scruffy research and make it neat. He even gave it a name: undertaking a "rational reconstruction" of some past research. In the late 1980s [6], the DReaM group embarked on a project to rationally reconstruct the rather scruffy inductive theorem proving techniques to be found in Boyer and Moore's NQTHM theorem prover [7]. Dieter Hutter in Karlsruhe was going about a similar task developing the INKA prover [8] and soon became closely involved with the efforts in Scotland.

Central to the inductive theorem proving heuristics in NQTHM were the ideas of recursion analysis (picking an induction rule and variable), and then rewriting the step case to match the induction hypothesis in a process that became rationally reconstructed as an annotated form of rewriting called "rippling" [9, 10]. Picking an appropriate induction rule and variable depends on how you can simplify the resulting step case so the two ideas are closely connected.

In the early 1990s, rippling ran into some annoying problems. In particular, the annotations used by rippling to guide rewriting could become ill-formed, and there was as yet no principled way to annotate terms in the first place. Alan's newest postdoc, David Basin, and I set about fixing these problems. I shall tell the story backwards as it makes for a more rational reconstruction of the work.

3.2.1 A Calculus for Rippling

Rippling is a form of rewriting guided by special kinds of (meta-level) annotation. Consider, for example, a proof that appending a list onto the empty list leaves the list unchanged. In the step case, the induction hypothesis is

$$(append\ x\ nil) = x.$$

From this, we need to derive the induction conclusion,

$$(append\ (cons\ e\ x)\ nil) = (cons\ e\ x).$$

We can annotate the induction conclusion to highlight the differences between it and the induction hypothesis,

$$(append\ \boxed{(cons\ e\ \underline{x})}^{\uparrow}\ nil) = \boxed{(cons\ e\ \underline{x})}^{\uparrow}.$$

The square boxes are the "wavefronts". The underlined parts are the "waveholes". If we eliminate everything in the wavefronts but not in the waveholes, we get the "skeleton" in which the induction conclusion matches exactly the induction hypothesis.

To simplify the induction hypothesis, we will use a rewrite rule derived from the recursive definition of append

$$(append\ (cons\ a\ b)\ c) =_{\text{def}} (cons\ a\ (append\ b\ c)).$$

We can turn this into a rewrite rule annotated with wavefronts and waveholes,

$$(append\ \boxed{(cons\ a\ \underline{b})}^{\uparrow}\ c) \Rightarrow \boxed{(cons\ a\ \underline{(append\ b\ c)})}^{\uparrow}.$$

This is called a "wave rule". It preserves the skeleton *(append a c)* but moves the wavefront *(cons a …)* up and hopefully out of the way. Applying this rule to the left-hand side of the induction conclusion gives

$$\boxed{(cons\ e\ \underline{(append\ x\ nil)})}^{\uparrow} = \boxed{(cons\ e\ \underline{x})}^{\uparrow}.$$

We can now simplify the left-hand side using the induction hypothesis as the rewrite rule,

$$(append\ x\ nil) \Rightarrow x.$$

This rewriting step is called fertilization and leaves the following equation:

$$\boxed{(cons\ e\ \underline{x})}^{\uparrow} = \boxed{(cons\ e\ \underline{x})}^{\uparrow}.$$

The left-hand side of the rewritten induction conclusion now matches the right-hand side. The step case therefore holds by the definition of equality.

Rewriting annotated terms in this way takes us beyond the normal rewriting of terms. David and I therefore formalized a calculus for describing such annotated rewriting [11, 12]. We showed that this calculus had the following four desirable properties.

Well-formedness: Rewriting properly annotated terms with wave rules leaves them properly annotated.

Skeleton preservation: Rewriting properly annotated terms with wave rules preserves the skeleton.

Correctness: We can perform the corresponding derivation in the underlying unannotated theory. Annotation is thus merely guiding search.

Termination: Given the appropriate measures on annotated terms, we can guarantee rippling terminates. There are, for example, no loops.

We showed that different termination orderings can profitably used within and outwith induction [13]. Such new orderings let us combine the highly goal-directed features of rippling with the flexibility and uniformity of more conventional term rewriting. For instance, we proposed two new orderings that allow unblocking, definition unfolding, and mutual recursion to be added to rippling in a principled (and terminating) fashion.

3.2.2 *Difference Matching and Unification*

But where do the annotations come from in the first place? David and I generalized both (1-sided) matching and (2-sided) unification to annotate terms appropriately. Difference matching extended first-order matching to make one term, the pattern match another, the target by instantiating variables in the target, as well as by hiding structure with wavefronts also in the target [14]. Difference unification extended unification to make two terms syntactically equal by variable instantiation and by hiding structure with wavefronts in both terms [15]. Difference unification is needed, for instance, to annotate rewrite rules as wave rules.

A single difference match can be found in time linear in the size of the target. If the pattern contains a variable, then set this to the target and put everything else in the wavefront. If not, the pattern is ground and we can simply descend through the term structure hiding any differences between the pattern and the target in wavefronts. However, there can be exponentially many difference matches in the size of the pattern in general so returning all of them can take exponential time. In practice, though, there are usually only a few successful difference matches and these can be found quickly.

Difference unification is more problematic computationally. Even if we limit wavefronts to one term, deciding if two terms difference unify together is NP-hard (Theorem 8 in [15]). Thus, supposing P $\neq$ NP, we cannot in general find even a single difference unifier in polynomial time. Looking again at this result more than 25 years later, I would not leave the analysis there but would look closer at the source of complexity. Difference unification has not proved too intractable in practice and we can likely argue why not. The reduction showing NP-hardness reduced a propositional satisfiability problem in m clauses to difference unifying two m-ary functions. The functions being difference unified in this proof therefore can have very great arity. I conjecture that difference unification is polynomial, more precisely fixed-parameter tractable, when applied to terms of bounded arity.

An unexpected tale:
To find the difference unification with the least amount of annotation, we proposed a new generic AI search called left-first search (LFS) [15]. Left branches of our search tree introduced annotations, whilst right branches matched terms. Left-first search explored leaf nodes of this search tree in order of the number of left branches taken. I presented the search method at IJCAI 1993.

Two years later, I was listening to an IJCAI 1995 conference talk on a new search method called limited discrepancy search (LDS) [16] when one of the authors put up a slide showing the order of the leaf nodes explored by LDS. This appeared identical to that of LFS, a slide I remember preparing 2 years before. At the end of the talk, I therefore asked about the difference between the two search methods. A colleague called it the "question from hell", but my intention was just to understand how they differed.

Unbeknown to me, LDS was being patented, and it set off a chain of unfortunate events. Lawyers had to rewrite the patent application at some significant cost. I was asked to be an expert witness in a patent dispute over LDS. And I was considered by the authors of LDS to be a "trouble maker".

Eventually, it would blow over as there is a simple but crucial difference. Our search trees were small and so LFS expanded them in memory. LDS was intended for much larger search trees and so, whilst it expanded leaf nodes in the same order as LFS, did so in a lazy space efficient fashion by returning repeatedly to the root node much like iterative deepening search. This adds just a constant factor to the time asymptotically so is worth paying when space is an issue.

Whilst difference unification was invented to deal with inductive proof, it captures a deeper and more general idea used in mathematics. In [17], J.A. Robinson presented a simple account of unification in terms of difference reduction. He observed,

> *"Unifiers remove differences ... We repeatedly reduce the difference between the two given expressions by applying to them an arbitrary reduction of the difference and accumulate the product of these reductions. This process eventually halts when the difference is no longer negotiable [reducible via an assignment], at which point the outcome depends on whether the difference is empty or nonempty".*

Difference unification can be seen as a direct extension of Robinson's notion of difference reduction: we reduce differences not just by variable assignment, but also by term structure annotation. However, what makes this extended notion of unification attractive is that this annotation is precisely what is required for rippling to remove this difference. And, as we see shortly, rippling has found a useful role to play in a number of other proof areas.

3.3 Proof Planning

An important idea explored within the DReaM group is the separation of logic and control. Proof planning was originally developed for inductive proof [18]. It brought together ideas of meta-level control explored in the earlier PRESS project [19] with AI planning operators specified by pre- and post-conditions. Theorem proving heuristics are described by general purpose proof planning methods such as rippling and fertilization that are glued together using simple AI search techniques like depth-first or best-first search. Since proof planning was proving useful for inductive proof, I became keen to try to apply it elsewhere.

3.3.1 Summing Series

To explore the use of proof planning in general, and rippling in particular outside of inductive proof, I chanced on the domain of summing series [20]. Inductive proofs can be used to verify identities about finite sums. But where do these identities come from in the first place?

I developed a set of proof planning methods to solve such problems. To my surprise, rippling proved to be key to many of these methods. I will illustrate this with the CONJUGATE method. This method transforms a finite sum of terms into the finite sum of some conjugates. The conjugate can be one of several second-order operations, e.g., the differential or integral of the original term, or the mapping of a trigonometric series onto the real or imaginary part of a complex series.

Consider, for example, finding a closed-form expression for

$$\sum_{i=0}^{n}(i+1)x^{i}.$$

The CONJUGATE method transformed this into a simpler looking sum,

$$\sum_{i=0}^{n}\frac{\mathrm{d}x^{i+1}}{\mathrm{d}x}.$$

This now looks close to a known result, the closed-form sum of a geometric series,

$$\sum_{i=0}^{n} x^i = \frac{x^{n+1} - 1}{x - 1}.$$

Difference matching our simpler looking sum against the left-hand side of this known result gives some wavefronts we need to remove out of the way by rippling with wave rules,

$$\sum_{i=0}^{n} \boxed{\frac{d\, x^{\boxed{i+1}^{\uparrow}}}{dx}}^{\uparrow}.$$

Since the derivative of a sum is the sum of the derivatives, rippling gives

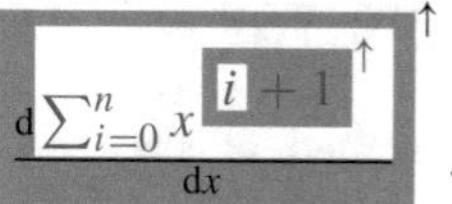

Rippling with a wave rule derived from the definition of exponentiation then expands the exponent,

$$\boxed{\frac{d \sum_{i=0}^{n} \boxed{x . x^i}^{\uparrow}}{dx}}^{\uparrow}.$$

One final rewriting step uses rippling to move the constant term outside the sum,

$$\boxed{\frac{d\, x . \sum_{i=0}^{n} x^i}{dx}}^{\uparrow}.$$

The FERTILIZE method substitutes the closed-form sum for the geometric series,

$$\boxed{\frac{d\, x . \frac{x^{n+1}-1}{x-1}}{dx}}^{\uparrow}.$$

Finally, a DIFFERENTIATE method then symbolically computes a closed-form answer by algebraically differentiating the quotient. The derivation is now complete.

We subsequently looked at some other mathematical domains such as theorems about limits [21]. Rippling and proof planning again proved up to the challenge.

3.3.2 A Divergence Critic

Proof planning methods come with high expectations of success. Their failure can therefore be a useful tool in patching proofs. I explored how the failure of rippling can be used to suggest missing lemma needed to complete a proof by means of a "divergence critic" [22, 23]. Other members of the DReaM group have explored similar ideas in closely related settings (e.g., [24, 25]).

My divergence critic identified when a proof attempt is diverging by means of difference matching. The critic then proposed lemmas and generalizations of these lemmas to try to allow the proof to go through without divergence. For example, when the prover failed to show inductively that $(rev\ (rev\ x)) = x$, the critic proposed the key lemma, a missing wave rule needed to complete the proof,

$$(rev\ \boxed{(append\ \underline{X}\ (cons\ Y\ nil))}^{\uparrow}) = \boxed{(cons\ Y\ \underline{(rev\ X)})}^{\uparrow}.$$

In my view, such failure is something we still exploit too little in automating mathematical reasoning. As a mathematician, I spend most of my time failing to prove conjectures. But those failures eventually often lead me to find either a proof when the conjecture is true, or a counter-example when it is false.

3.4 Mathematical Discovery

Mathematics is more than just proving theorems. It is also defining theories, inventing definitions, conjecturing results, finding counter-examples, developing proof methods, and more. One of the pleasures of the DReaM group was to witness and contribute to automating some of these other mathematical activities.

In 1996, Alan started to supervise a young and ambitious PhD student, Simon Colton. I was lucky enough to help out. Simon wanted to build a system to invent new theories. Doug Lenat had shown the feasibility of doing this with the AM and followup Eurisko systems [26, 27]. Simon set out to rationally reconstruct Lenat's work in his HR program [28–31]. This was named appropriately after the famous double act, Hardy and Ramanujan. Actually, HR was not much of a rational reconstruction of AM other than to work on the same problem as AM, and to use a two letter name like AM.

In a wonderful example of why PhD students should not listen to their supervisors, I suggested to Simon to keep well away from number theory. I reasoned that there had been thousands of years of attention to number theory. New and automated mathematical discoveries were more likely therefore to be found in some newer and little studied theory like that of Moufang loops. Fortunately, Simon ignored this advice and HR made a number of discoveries in number theory. On the other hand, in a wonderful example of why PhD students should listen to their supervisors, Simon was not keen to submit an update on his work to AAAI 2000. I persuaded him to do so, and the paper won the Best Paper award.

3.5 The Meta-Level

One of the other rewards of working in the DReaM group was Alan's attention to the meta-level. This was not just the meta-mathematical level, but the meta-level of doing research. Alan thought long and hard about how we do research, and how you could do it better. I still recommend the Researcher's Bible that Alan co-wrote to my PhD students whether they were starting out, or writing up their thesis [32]. And when I left Edinburgh, I "borrowed" many of his techniques for doing research on my own: writing half-formed ideas down in internal notes, trying to think of questions to ask at every seminar, giving informal talks on any interesting papers I had seen at summer conferences, etc.

3.6 Conclusions

Out of interest, I downloaded one of Alan's latest paper [33]. To my surprise and pleasure, it repeats and expands on many of the ideas I have discussed here. It applies rippling to a new domain, invariant preservation proofs. The meta-level guidance rippling provides is used to build proof patches to recover failed attempt and eventually finish the proofs. And the paper ends with an appendix containing a formal definition of rippling, along the lines of the calculus we presented 25 years ago. It feels just like yesterday. Thank you for everything, Alan.

Acknowledgments Funded by the European Research Council under the Horizon 2020 Programme via the Advanced Research grant AMPLify 670077.

References

1. Walsh, T.: A Theory of Abstraction. PhD thesis, University of Edinburgh (1991)
2. Giunchiglia, F., Walsh, T.: A Theory of Abstraction. Artificial Intelligence **56**(2–3) (1992) 323–390 Also available as DAI Research Paper No 516, Dept. of Artificial Intelligence, Edinburgh.
3. Giunchiglia, F., Villafiorita, A., Walsh, T.: Theories of abstraction. AI Communications **10**(3,4) (1997) 167–176
4. Rossi, F., van Beek, P., Walsh, T., eds.: Handbook of Constraint Programming. Foundations of Artificial Intelligence. Elsevier (2006)
5. Biere, A., Heule, M., van Maaren, H., Walsh, T., eds.: Handbook of Satisfiability. Frontiers in Artificial Intelligence and Applications. IOS Press (2009)
6. Bundy, A., Van Harmelen, F., Hesketh, J., Smaill, A., Stevens, A.: A rational reconstruction and extension of recursion analysis. In: Proceedings of the 11th International Joint Conference on Artificial Intelligence. IJCAI'89, San Francisco, CA, USA, Morgan Kaufmann Publishers Inc. (1989) 359–365
7. Boyer, R., Moore, J.: A Computational Logic. Academic Press (1979) ACM monograph series.
8. Hutter, D.: Guiding inductive proofs. In Stickel, M., ed.: 10th International Conference on Automated Deduction, Springer-Verlag (1990) 147–161 Lecture Notes in Artificial Intelligence No. 449.

9. Bundy, A., van Harmelen, F., Smaill, A., Ireland, A.: Extensions to the rippling-out tactic for guiding inductive proofs. In Stickel, M., ed.: 10th International Conference on Automated Deduction, Springer-Verlag (1990) 132–146 Lecture Notes in Artificial Intelligence No. 449. Also available from Edinburgh as DAI Research Paper 459.
10. Bundy, A., Stevens, A., van Harmelen, F., Ireland, A., Smaill, A.: Rippling: A heuristic for guiding inductive proofs. Artificial Intelligence **62** (1993) 185–253 Also available from Edinburgh as DAI Research Paper No. 567.
11. Basin, D., Walsh, T.: A calculus for rippling. In: Proceedings of CTRS-94. (1994)
12. Basin, D., Walsh, T.: A calculus for and termination of rippling. Journal of Automated Reasoning **16**(1–2) (1996) 147–180
13. Basin, D., Walsh, T.: Termination orderings for rippling. In Bundy, A., ed.: 12th Conference on Automated Deduction, Springer Verlag (1994) 466–483 Lecture Notes in Artificial Intelligence No. 814.
14. Basin, D., Walsh, T.: Difference matching. In Kapur, D., ed.: 11th Conference on Automated Deduction, Springer Verlag (1992) 295–309 Lecture Notes in Computer Science No. 607. Also available from Edinburgh as DAI Research Paper 556.
15. Basin, D., Walsh, T.: Difference unification. In: Proceedings of the 13th IJCAI, International Joint Conference on Artificial Intelligence (1993)
16. Harvey, W.D., Ginsberg, M.L.: Limited discrepancy search. In: Proceedings of the 14th IJCAI, International Joint Conference on Artificial Intelligence (1995) 607–613
17. Robinson, J.: Notes on resolution. In Bauer, F., ed.: Logic, Algebra, and Computation. Springer Verlag (1989) 109–151
18. Bundy, A.: The use of explicit plans to guide inductive proofs. In Lusk, E.L., Overbeek, R.A., eds.: 9th International Conference on Automated Deduction. Volume 310 of Lecture Notes in Computer Science., Springer (1988) 111–120
19. Sterling, L., Bundy, A., Byrd, L., O'Keefe, R.A., Silver, B.: Solving symbolic equations with PRESS. J. Symb. Comput. **7**(1) (1989) 71–84
20. Walsh, T., Nunes, A., Bundy, A.: The use of proof plans to sum series. In Kapur, D., ed.: 11th International Conference on Automated Deduction. Volume 607 of Lecture Notes in Computer Science., Springer (1992) 325–339
21. Yoshida, T., Bundy, A., Green, I., Walsh, T., Basin, D.A.: Coloured rippling: An extension of a theorem proving heuristic. In Cohn, A.G., ed.: Proceedings of the Eleventh European Conference on Artificial Intelligence, John Wiley and Sons, Chichester (1994) 85–89
22. Walsh, T.: A divergence critic. In Bundy, A., ed.: 12th Conference on Automated Deduction, Springer Verlag (1994) 14–25 Lecture Notes in Artificial Intelligence No. 814.
23. Walsh, T.: A divergence critic for inductive proof. Journal of Artificial Intelligence Research **4** (1996) 209–235
24. Ireland, A., Bundy, A.: Extensions to a generalization critic for inductive proof. In McRobbie, M.A., Slaney, J.K., eds.: 13th International Conference on Automated Deduction. Volume 1104 of Lecture Notes in Computer Science., Springer (1996) 47–61
25. Ireland, A.: Productive use of failure in inductive proof. J. Autom. Reasoning **16**(1-2) (1996) 79–111
26. Lenat, D.B., Brown, J.S.: Why AM and Eurisko appear to work. In Genesereth, M.R., ed.: Proceedings of the National Conference on Artificial Intelligence (AAAI83), AAAI Press (1983) 236–240
27. Lenat, D.B., Brown, J.S.: Why AM and EURISKO appear to work. Artif. Intell. **23**(3) (1984) 269–294
28. Colton, S., Bundy, A., Walsh, T.: Automatic identification of mathematical concepts. In: Proceedings of 16th IJCAI, International Joint Conference on Artificial Intelligence (1999)
29. Colton, S., Bundy, A., Walsh, T.: Automatic invention of integer sequences. In: Proceedings of the 16th National Conference on AI, Association for Advancement of Artificial Intelligence (2000) 558–563
30. Colton, S., Bundy, A., Walsh, T.: Automatic identification of mathematical concepts. In: Proceedings of the 17th International Conference on Machine Learning. (2000)

31. Colton, S., Bundy, A., Walsh, T.: On the notion of interestingness in automated mathematical discovery. International Journal of Human-Computer Studies **53**(3) (2000) 351–375
32. Bundy, A., Du Boulay, B., Howe, J., Plotkin, G.: The researchers' bible. Department of Artificial Intelligence, University of Edinburgh (1985)
33. Lin, Y., Bundy, A., Grov, G., Maclean, E.: Automating Event-B invariant proofs by rippling and proof patching. Formal Asp. Comput. **31**(1) (2019) 95–129

Chapter 4
Dynamic Proof Presentation

Paul B. Jackson

Abstract For several decades, there has been significant debate over the formal proof style supported by proof assistants. For example, the merits of a declarative style rather than a procedural (tactic) style have been argued. In much of the debate, there has been unnecessarily rigid insistence on the languages of proof input and proof presentation being identified. When these concepts are not shackled together, many opportunities are opened up for dynamic proof presentation that take full advantage of the capabilities of computer user interfaces. With dynamic proof presentation, the proof viewer can easily focus attention on particular parts of proofs and change the level of detail presented. One viewer might be interested in just a proof outline, another might want to see how a large step of inference is composed of smaller steps. Current proof assistant user interfaces do provide some dynamic presentation capabilities, but much more could be done. Further attention to dynamic proof presentation should help make formal proofs easier to understand by a wider range of audiences, with minimal need to rewrite proof libraries that are developed with huge time investments.

4.1 Introduction

4.1.1 Proof Presentation Style

The core topic this chapter addresses is that of how formal proofs created using interactive theorem provers ought to be presented. Specifically, the concern is with the presentation of the structure of proofs of individual lemmas, rather than the presentation of theories grouping lemmas and definitions, or the presentation of terms, types and formulas.

P. B. Jackson (✉)
University of Edinburgh, Edinburgh, UK
e-mail: Paul.Jackson@ed.ac.uk

G. Michaelson (eds.), *Mathematical Reasoning: The History and Impact of the DReaM Group*, https://doi.org/10.1007/978-3-030-77879-8_4

Over recent decades, there has been significant discussion of this topic [9, 12, 25, 27]. It has long been recognised that tactic scripts by themselves are poor at communicating proof structure to readers. Tactic scripts describe how the prover should create or check full proofs, but typically do not show intermediate subgoal formulas and often obscure the branching structure of proofs. Often, if a reader is to understand a proof, they need access to a working version of the prover and they have to re-execute the tactic script step by step.

Proofs in declarative proof description languages (e.g., the Mizar system language and the Isar language of the Isabelle prover) are generally much more readable than tactic scripts. With such languages, the proof text explicitly states many of the intermediate formulas in a proof, nested-blocks describe hierarchical proof structure and the syntax is designed to be reminiscent of that found in mathematics papers and textbooks. These language features help a reader quickly gain a level of understanding of a proof just from study of its formal text, without running the relevant prover.

4.1.2 Ongoing Issues

The tactic style (sometimes called the *procedural style*) is still the norm in the user communities of a number of theorem provers (those for Coq, HOL4 and HOL Light, for example). A disincentive for tactic users to move to a more declarative style is that the intermediate formulas needed for a declarative style can be tedious to enter and can make the proof scripts considerably more verbose. This is all the more the case when the formulas get large, as is common in formal verification applications. The lower comprehensibility of tactic-style proofs is not so much an issue when proofs are developed using tactics, as then the prover user interfaces present sufficient subgoal information to orient developers.

A key limitation of current practices of writing both declarative and tactic-style proofs is that the level of detail is fixed at proof writing time. Sometimes this level is determined by the extent of automation provided by the prover. Some arguments might need to be spelled out in more detail than a reader might want. Other times automation might enable large steps whose details are non-obvious to the reader.

In general, there will be no one optimal level of detail. Different readers might have widely different degrees of familiarity with the prover and the subject formalised. And readers have different interests in proofs at different times. Sometimes readers are keen to understand the details; perhaps they wish to reproduce proofs in another prover or perhaps they are studying the proofs for a mathematics or computer science class. Other times, maybe they just want a high-level summary.

As mentioned above, replay of tactic-style proofs is usually essential for gaining a good understanding of them. Replay too is often useful even for declarative proofs, as some details of intermediate proof formulas are only viewable on replay. This need for replay is a significant barrier to any reader who does not wish to go to the trouble of installing the relevant prover and learning the basics of its usage. Further, it might take minutes or even hours to get a prover into a state where some given proof of interest can be replayed.

Additional issues concerning readability include that proof intelligibility varies significantly depending on the proof writer and that there is a huge body of existing proofs, many written in a procedural style, to which it would be good to have better access.

4.1.3 The Vision

This chapter describes how to improve support for dynamically choosing levels of proof presentation detail. The ideas discussed are relevant to proofs in both declarative and procedural styles, and they could make the differences between these styles less significant. Nearly all the ideas have already been experimented with in some way, so their further development and integration should be relatively straightforward. This improved support could significantly ease and speed the understanding of formal proofs.

Interest in using interactive theorem provers has been steadily increasing, both from those wishing to use them for formal verification and from those exploring their use for education and research in mathematics and computer science. Dynamic proof presentation technologies would be much appreciated by many of these users and could spur on further growth in theorem prover user communities.

4.1.4 Structure of Rest of Chapter

Section 4.1.5 includes a few notes on my interactions with the DReaM group over the years. Sections 4.2 to 4.6 sketch out what dynamic proof presentation can look like when both procedural and declarative proof styles are used. A number of DReaM group members have worked on relevant topics, and Sect. 4.7 describes this work. Sections 4.8 and 4.9 are forward looking, considering desired technical requirements for supporting dynamic proof presentation and the relationship between proof viewing and proof editing. Further related work by the DReaM group associates and others is considered in Sect. 4.10, and finally, Sect. 4.11 summarises work strands that could be profitable in future.

4.1.5 My DReaM Group Connections

I came to the University of Edinburgh in 1995, having recently completed a PhD at Cornell University with Bob Constable on enhancing the Nuprl interactive theorem prover and using it for formalising some abstract algebra. Constable had previously visited Edinburgh on sabbatical, and the DReaM group's Oyster interactive theorem prover developed in the late 1980s was strongly inspired by Nuprl. Initially, I had a post-doc in the LFCS (the Laboratory for Foundations of Computer Science) with Rod Burstall and then, from 1998, a lectureship.

My PhD work gave me a keen interest in the central concerns of the group about the automation of mathematical reasoning. Over the years, I have enjoyed very much attending and participating in the DReaM group talk meetings, and I continue to do so now. I appreciate the informality of these meetings; there can be as much or even more time spent in lively discussion as spent by the speaker talking. It is rare that they are like a regular seminar when there might be just a couple of polite questions afterwards. More generally, the group for me has been a welcoming intellectual home.

From 1998 to 2019, I was a co-investigator on grants held by the group that funded foundational and pump-priming research. For the most part, my research has followed paths closely related to but distinct from those of other group members; topics I have pursued include bounded model checking, the formal verification of software, automatic theorem proving for non-linear arithmetic and, most recently, the verification of hybrid dynamical systems. This last topic is now also of interest to DReaM group member Jacques Fleuriot, and we are currently planning a collaboration in this area.

The issue of how formal proofs should be presented has been an interest of mine ever since starting to work with Nuprl in the late 1980s. A couple of times I was involved in discussions with DReaM group members (primarily Alan Bundy and David Aspinall) about pursuing funding related to the topic of this chapter. Unfortunately, neither time did we develop ideas to the stage of completing and submitting a funding proposal. Recently, I have been enthusiastic about the rise in prominence of the Lean theorem prover [17]. It has a rapidly growing formal library and has attracted significant interest from mathematicians in using it in both their teaching and their research. This has renewed my interest in formal proof presentation, and I might well use Lean for future work in this area.

4.2 A Running Example of a Procedural Proof

To help motivate the discussion throughout this chapter, let us use a proof of a lemma from the Nuprl system [6] that is the penultimate step in a proof of the irrationality of $\sqrt{2}$. This proof was previously presented in a comparison of 17 theorem provers [26], and it follows the shape of a proof example used by Lamport when advocating a structured proof format [14].

Figure 4.1 shows an automatically generated *tactic-and-subgoal* proof-tree presentation of the proof. Nuprl is an interactive theorem prover in the LCF family: a proof of a lemma is undertaken by running tactics—procedural proof commands—on goals. Goals are sequents with numbered variable declarations and hypotheses and a single conclusion. When a tactic is run on a goal, 0 or more subgoals result.

```
*T root_2_irrat_over_int

⊢ ¬(∃m,n:ℤ. CoPrime(m,n) ∧ m * m = 2 * n * n)
|
BY (D 0 THENM ExRepD ...a)
|
1. m: ℤ
2. n: ℤ
3. CoPrime(m,n)
4. m * m = 2 * n * n
⊢ False
|
BY Assert ⌜2 | m⌝
|\
| ⊢ 2 | m
| |
| BY (BLemma 'two_div_square' THENM Unfold 'divides' 0
|     THENM AutoInstConcl [] ...a)
 \
  5. 2 | m
  |
  BY Assert ⌜2 | n⌝
  |\
  | ⊢ 2 | n
  | |
  | BY (BLemma 'two_div_square' THENM All (Unfold 'divides')
  |     THENM ExRepD THENM Inst [⌜c * c⌝] 0
  |     THENM RWO "6" 4 ...)
   \
    6. 2 | n
    |
    BY (RWO "coprime_elim" 3 THENM FHyp 3 [5;6] ...a)
    |
    3. ∀c:ℤ. c | m ⇒ c | n ⇒ c ~ 1
    7. 2 ~ 1
    |
    BY (RWO "assoced_elim" 7 THENM D (-1) ...)
```

Fig. 4.1 Tactic-and-subgoal proof-tree presentation

If no subgoals result, then the tactic completely proves the goal. Otherwise, the goal the tactic is run on is only proven once the subgoals are completely proven by further tactic runs. As tactics can be combined using tacticals into larger tactics, it is always possible to compose a single tactic that completely proves a top-level goal. Traditionally, this is only done in Nuprl when the proof is very straightforward. Otherwise, proofs are presented in a tree form, with goal nodes and tactic nodes alternating as one moves down the tree branches. That is what one sees in Fig. 4.1: a tree of goals, separated by tactic calls after occurrences of the `BY` keyword.

To save space, this proof-tree presentation elides the repetition of variable declarations, hypotheses and conclusions in the goal sequents. For example, Fig. 4.2

With elision

```
...
|
1. m: ℤ
2. n: ℤ
3. CoPrime(m,n)
4. m * m = 2 * n * n
⊢ False
|
BY Assert ⌈2 | m⌉
|\
| ⊢ 2 | m
| |
| ...
 \
  5. 2 | m
  |
  ...
```

Without elision

```
...
|
1. m: ℤ
2. n: ℤ
3. CoPrime(m,n)
4. m * m = 2 * n * n
⊢ False
|
BY Assert ⌈2 | m⌉
|\
| 1. m: ℤ
| 2. n: ℤ
| 3. CoPrime(m,n)
| 4. m * m = 2 * n * n
| ⊢ 2 | m
| |
| ...
 \
  1. m: ℤ
  2. n: ℤ
  3. CoPrime(m,n)
  4. m * m = 2 * n * n
  5. 2 | m
  ⊢ False
  |
  ...
```

Fig. 4.2 A proof step with and without elision of repeated sequent components

shows the second step of the Fig. 4.1 proof presentation, with and without elision of repeated sequent components. To construct a complete picture of some sequent in this kind of proof presentation, the reader needs to search up the tree for any elided components. It helps to know that proof steps in Nuprl might change existing components of the declaration and hypothesis list, or add new components, or both, but it is rare that components are deleted and a subgoal after a step has a shorter list. In the event that a subgoal's declaration and hypothesis list is shorter than that of the parent goal, there is no elision of components in the subgoal, in order to avoid ambiguity.

In the Fig. 4.1 proof, the first proof step shows that the proof strategy is to assume the negation of the goal, and from this to show falsity, i.e., that we have a contradiction. The tree presentation makes clear that the proof proceeds by first establishing 2 divides m (hypothesis 5) and then that 2 divides n (hypothesis 6). The contradiction then follows because hypothesis 3 claims that m and n are co-prime, that they have no non-trivial common divisors. A divisor is trivial if it is a unit as far as divisibility is concerned, i.e., if it is $+1$ or -1. We also see two side proofs in the tree presentation: the first establishing that 2 divides m and the second, knowing that 2 divides m, that also 2 divides n.

```
*A divides          b | a ==  ∃c:ℤ. a = b * c
*A assoced          a ~ b ==  a | b ∧ b | a
*A gcd_p            GCD(a;b;y) ==  y | a ∧ y | b ∧
                                   (∀z:ℤ. z | a ∧ z | b ⇒ z | y)
*A coprime          CoPrime(a,b) ==  GCD(a;b;1)
*T two_div_square   ∀n:ℤ. 2 | n * n ⇒ 2 | n
*T coprime_elim     ∀a,b:ℤ.  CoPrime(a,b) ⟺
                             (∀c:ℤ. c | a ⇒ c | b ⇒ c ~ 1)
*T assoced_elim     ∀a,b:ℤ.  a ~ b ⟺ a = b ∨ a = -b
```

Fig. 4.3 Definitions and lemmas

What do the tactics in Fig. 4.1 actually do? Nuprl's tactic language, and indeed most procedural proof languages, require study to understand. Sometimes names are suggestive (e.g., `Assert` or `Unfold`); other times they are rather abbreviated to save space and typing (e.g., `RWO` is short for *rewrite once*). Also lemma names (e.g., `coprime_elim`, `assoced_elim`) and definition names (e.g., `divides` for the infix | operator) provide part of the story. Once the reader sees the referenced lemmas and the definitions used, they can sometimes make a fair guess as to what is going on. For the lemmas and definitions relevant to the running example proof, see the fragment of a Nuprl library listing in Fig. 4.3. Here, lines starting with `*A` are for definitions (`A` is for *abstraction*, Nuprl's terminology for a definition) and with `*T` are for lemmas and theorems. Note that, when doing divisibility theory over the integers, GCDs are unique only up to associates (as specified by the `assoced` relation, with ~ infix notation).

The next sections explore how dynamic proof presentation capabilities can improve proof readability and understandability.

4.3 Focussing on Proof Steps in Procedural Proofs

While a tactic-and-subgoal proof tree as in Fig. 4.1 can provide a good overview of a proof, such trees become hard to read when spread over many pages as the vertical linearisation creates distance between a goal and its immediate subgoals. If our attention is on some subtree, it is easy to have the presentation start at the root goal of that subtree rather than the initial goal being proven. If the tree presentation is in an interactive viewer, a facility for hiding subtrees that are not of immediate interest is useful. If our attention is on a particular goal, showing just that goal and its immediate subgoals is helpful. In Nuprl, such a view is the primary way proofs are interactively presented, both when viewing proofs and editing proofs. For example, Fig. 4.4 shows the view Nuprl would give of the step introducing the `2 | n` hypothesis 6. As with the full tactic-and-subgoal proof trees, subgoals omit repeated sequent components to save space and enable the reader to focus on what has changed. A `*` at the start of a sequent indicates that the proof below that point is complete. If the proof is incomplete, a `#` is used instead. The `top 1 2` is the

```
* top 1 2
1. m: ℤ
2. n: ℤ
3. CoPrime(m,n)
4. m * m = 2 * n * n
5. 2 | m
⊢ False

BY Assert ⌈2 | n⌉

1* ⊢ 2 | n

2* 6. 2 | n
     ⊢ False
```

Fig. 4.4 A proof refinement step

tree address of the top sequent. Users navigate up and down the proof tree just by clicking on the goal or one of the subgoals in such a view.

4.4 Condensing Tactic-and-Subgoal Proof Trees

The length of a proof-tree presentation can be reduced by combining adjacent tactics with tacticals. For example, the two steps at the end of the presentation in Fig. 4.1

```
6. 2 | n
|
BY (RWO "coprime_elim" 3 THENM FHyp 3 [5;6] ...a)
|
3. ∀c:ℤ. c | m ⇒ c | n ⇒ c ~ 1
7. 2 ~ 1
|
BY (RWO "assoced_elim" 7 THENM D (-1) ...)
```

could be collapsed into the single step:

```
   6. 2 | n
|
BY (RWO "coprime_elim" 3 THENM FHyp 3 [5;6]
    THENM RWO "assoced_elim" 7 THENM D (-1) ...)
```

The reader might be wondering what the "..." and "...a" signify in the Nuprl tactics shown here and earlier. These are notational shorthand for calls of Nuprl's auto-tactic on resulting subgoals. This tactic undertakes common straightforward reasoning steps such as proving linear arithmetic facts and checking well-formedness of terms and formulas. (In Nuprl, type checking is undecidable, and all type checking is undertaken using proof.) The "..." variant is for running the auto-tactic on all subgoals, and the "...a" variant is for running the auto-tactic only on *auxiliary* subgoals such as well-formedness subgoals.

Collapsing tactic steps together usually decreases hints to the reader as to what is going on with each step, as intermediate goals are then no longer visible. It therefore can be helpful if the proof developer can add comments explaining steps. Indeed, a useful option is to hide the tactic text when comments are used. This can produce readable proof outlines that are accessible to those unfamiliar with the tactic language. Figure 4.5 shows what an outline of the whole running proof could look like if some adjacent steps are combined, comments are inserted and tactics are hidden.

Another possible viewing option could involve the replacement of tactic text with automatically generated natural-language explanations of the tactics. One simple way to realise this would be to associate every tactic with some natural-language description template with slots for appropriate printing of any tactic arguments. See Fig. 4.6 for a mock-up of how the running proof might look with such an approach. While this might be more accessible to a reader not familiar with Nuprl, it still assumes familiarity with concepts such as forward chaining, back chaining and rewriting, and the reader needs to understand that to *decompose* a hypothesis or conclusion is to apply some relevant left or right introduction rule in a backwards fashion.

There have been more sophisticated investigations of how to produce natural-language versions of whole tactic-based proofs. For example, see the work of Holland-Minkley on presenting Nuprl proofs [10]. Even if easily understandable renditions of tactic text can be automatically generated, there still is a need for supporting display of human-written comments, as these comments might provide higher-level motivation for *why* a proof is being steered some particular way.

```
*T root_2_irrat_over_int

⊢ ¬(∃m,n:ℤ. CoPrime(m,n) ∧ m * m = 2 * n * n)
|
BY Assume negation of goal and aim for proof by contradiction
|
1. m: ℤ
2. n: ℤ
3. CoPrime(m,n)
4. m * m = 2 * n * n
⊢ False
|
BY From hyp 4, deduce that 2 | m
|
5. 2 | m
|
BY From hyps 4 and 5, deduce that 2 | n
|
6. 2 | n
|
BY Observe that hyps 5 and 6 contradict hyp 3
```

Fig. 4.5 Proof outline

```
*T root_2_irrat_over_int

⊢ ¬(∃m,n:ℤ. CoPrime(m,n) ∧ m * m = 2 * n * n)
|
BY Decompose the conclusion
|  THEN Repeatedly decompose hypotheses,
|    including existential quantifiers
|
1. m: ℤ
2. n: ℤ
3. CoPrime(m,n)
4. m * m = 2 * n * n
⊢ False
|
BY Assert ⌈2 | m⌉
|\
| ⊢ 2 | m
| |
| BY Back-chain using the lemma two_div_square
|     THEN Unfold the definition of divides (|) in the conclusion
|     THEN Instantiate the conclusion's existential quantifier
|       by matching the quantifier body against some hypothesis
 \
  5. 2 | m
  |
  BY Assert ⌈2 | n⌉
  |\
  | ⊢ 2 | n
  | |
  | BY Back-chain using the lemma two_div_square
  |     THEN Unfold the definition of divides (|)
  |       in all hypotheses and the conclusion
  |     THEN Repeatedly decompose hypotheses,
  |       including existential quantifiers
  |     THEN Instantiate the conclusion's quantifier(s)
  |       with the term(s) ⌈c * c⌉
  |     THEN Rewrite hypothesis 4 using hypothesis 6
   \
    6. 2 | n
    |
    BY Rewrite hypothesis 3 using the lemma coprime_elim
    |  THEN Forward-chain using hypothesis 3,
    |    matching with hypotheses 5 and 6
    |
    3. ∀c:ℤ. c | m ⇒ c | n ⇒ c ~ 1
    7. 2 ~ 1
    |
    BY Rewrite hypothesis 7 using lemma assoced_elim
       THEN Decompose the last hypothesis
       THEN Repeatedly apply straightforward reasoning techniques
```

Fig. 4.6 Proof-tree presentation with simple natural-language rendering of tactics

4.5 Expanding Proof Steps

Sometimes a proof reader wishes to explore a proof step in more detail. For example, they might want to split apart the tactic steps combined using the `THENM` sequencing tactical ("then on *main* subgoal") that are used to prove the `2 | n` goal. See Fig. 4.7 for a copy of the original tactic-and-subgoal proof-tree fragment followed by a proof tree for the expanded version of this fragment. Now the reader can see, in

Original proof:

```
1. m: ℤ
2. n: ℤ
3. CoPrime(m,n)
4. m * m = 2 * n * n
5. 2 | m
⊢ 2 | n
|
BY (BLemma 'two_div_square' THENM All (Unfold 'divides')
    THENM ExRepD THENM Inst [⌈c * c⌉] 0
    THENM RWO "6" 4 ...)
```

Expanded proof:

```
1. m: ℤ
2. n: ℤ
3. CoPrime(m,n)
4. m * m = 2 * n * n
5. 2 | m
⊢ 2 | n
|
BY (BLemma 'two_div_square' ...a)
|
⊢ 2 | n * n
|
BY All (Unfold 'divides')
|
5. ∃c:ℤ. m = 2 * c
⊢ ∃c:ℤ. n * n = 2 * c
|
BY ExRepD
|
5. c: ℤ
6. m = 2 * c
|
BY (Inst ⌈c * c⌉ 0 ...a)
|
⊢ n * n = 2 * c * c
|
BY (RWO "6" 4 ...)
```

Fig. 4.7 Original and expanded proof of `2 | n`

the penultimate step, the definition of this variable `c` that is used in the term `c * c` used to instantiate the existential quantifier in the conclusion.

Further expansion could be desirable for tactics such as the auto-tactic that are defined in terms of a number of simpler tactics. Expansion of the auto-tactic at the very end of the proof could show the linear integer arithmetic tactic used to prove the main goal and the type checking tactic used to prove various well-formedness goals that are a by-product of the rewriting of hypothesis 4 with hypothesis 6.

Nuprl happens to have some support for such expansion, as it stores the proof-tree fragments created by tactic runs, and a proof editor command enables the replacement of a tactic run by the resulting proof tree. Unfortunately, by default, these proof-tree fragments are at the primitive rule level, which is far too detailed to be of interest to almost all readers. To arrange that higher-level tactics can expand into lower-level tactics, the code doing the expansion needs access to the syntax of tactic expressions and tactic definitions. With Nuprl, these details are hidden away in the ML compiler's data structures and are not accessible to the ML runtime. This is a general issue one has to face whenever tactics are expressed directly in some programming language. It is avoided when the prover adopts a custom proof command language and ASTs for commands are readily available. I did experiment with specially defined tactics and tacticals that captured structural information about tactics and enabled incremental expansion of tactic runs into lower-level tactics. However, it was difficult to do this for all tactics and this facility never made it into the standard Nuprl release.

4.6 Dynamic Presentation of Declarative Proofs

A declarative version of our running example proof is shown in Fig. 4.8. This uses the Isabelle Isar declarative proof language but, within the ⌜⌝ quotes, keeps the previously used Nuprl notation for terms and formulas. It has been derived from a proof undertaken using the Isabelle 2020 system.

From a content point of view, this is not so different from our initial procedural proof-tree presentation in Fig. 4.1. The high-level flow of the proof with the successively introduced hypotheses `CoPrime(m,n)`, `m * m = 2 * n * n`, `2 | m`, `2 | n` and `2 ~ 1` is the same. A minor difference is in how hypotheses are referred to: here they have symbolic labels rather than numbers and the special name `this` is used to refer to an unlabelled immediately previous hypothesis. Nested `proof-qed` blocks capture the side proofs of several of the introduced hypotheses. The `from` phrases make clear how earlier assumptions and lemmas are used in immediately following proof steps. After the `proof` and `by` keywords are instances of *methods*, Isabelle's version of tactics. Because Isar proofs still involve invocations of procedural tactics to justify declared steps, they are sometimes referred to as being *semi-declarative*. In other more purely declarative systems such as Mizar, virtually all steps are either basic steps of propositional and predicate logic or involve a single implicitly invoked procedure.

```
theorem root_2_irrat_over_int:
  ⌜¬(∃m,n:ℤ. CoPrime(m,n) ∧ m * m = 2 * n * n)⌝
proof
  assume ⌜∃m,n:ℤ. CoPrime(m,n) ∧ m * m = 2 * n * n⌝
  from this obtain m n where cop: ⌜CoPrime(m,n)⌝
                         and eq: ⌜m * m = 2 * n * n⌝ by auto
  have tdm: ⌜2 | m⌝
  proof (rule two_div_square)
    from eq show ⌜2 | m * m⌝ by (unfold divides, simp)
  qed
  have tdn: ⌜2 | n⌝
  proof (rule two_div_square)
    show ⌜2 | n * n⌝
    proof (unfold divides)
      from tdm obtain c where ⌜m = 2 *c⌝ by (unfold divides, auto)
      from this eq have ⌜n * n = 2 * c * c⌝ by simp
      from this show ⌜∃k. n * n = 2 * k⌝ by simp
    qed
  qed
  have ta1: ⌜2 ∼ 1⌝
  proof -
    from cop coprime_elim have ⌜∀c. c | m ∧ c | n ⇒  c ∼ 1⌝
      by simp
    from this tdm tdn show ⌜2 ∼ 1⌝ by auto
  qed
  show ⌜False⌝
  proof -
    from ta1 assoc_elim have ⌜2 = 1 ∨ 2 = -1⌝ by simp
    from this show ⌜False⌝ by arith
  qed
qed
```

Fig. 4.8 Declarative proof

From a proof creation point of view, the difference between declarative proofs and proof-tree presentations of procedural proofs is usually much more radical, as all of the text in the declarative case has to be entered in the proof source file. The proof developer has not only to enter the various keywords defining the shape of the proof and suggesting what deductions depend on, but also enter all the intermediate formulas introduced in the proof. To some extent, this further work by the proof developer is moderated because, knowing the result of a proof step, automation can do more to figure out how to justify a step given hints. Also, with Isabelle, the jEdit proof editor has a command that generates formula text for case splits and inductions, when the text can get rather tedious to figure out by hand.

With Isabelle, some practices act against readability. For example, proof method text automatically generated by the Sledgehammer tool [4] often contains rather more detail than many proof readers care about. (Sledgehammer is an all-purpose tool that combines a variety of automatic reasoning engines such as SMT solvers and first-order automatic theorem provers.) And there are some conventions for referring to parts of subgoals resulting from inductions and case splits that, while easing typing, avoid entry of and therefore also presentation of the full formulas involved in the subgoals.

Virtually all the ideas for dynamic proof presentation make sense in this declarative context, and some support is available. For example, with Isabelle's jEdit, the user can click at any point in a declarative proof, and a separate window shows some subgoal and context information associated with that position. And jEdit does support folding of `proof-qed` blocks, so the viewer has some control over the level of detail. It would be straightforward to allow source text to include some marks indicating blocks to be folded by default, so say just some comments on what the blocks do are visible. Expansion of proof commands would probably take some work. It may be that showing some kind of command execution traces in auxiliary windows would be easier than generating source text versions with more-detailed proof text. Indeed, Isabelle currently allows execution traces for its `simp` rewriting method to be displayed.

4.7 DReaM Group Contributions

Several DReaM group researchers have been concerned with the issue of how best to present partial views of proof plans so that the reader easily sees relationships between plan parts and is not overloaded with detail. The first three subsections below survey relevant work by these researchers.

A key observation in this work was the importance of being able to view hierarchies of both subgoals and proof methods. Later work covered in Sects. 4.7.4 and 4.7.5 formalised a notion of proof trees with these two hierarchies and used this formalisation to help reason about proof transformations that could make proofs easier to understand.

4.7.1 Barnacle and XBarnacle

Lowe, McLean and Bundy developed the Barnacle [15] and XBarnacle [16] graphical user interfaces (GUIs) to the CLAM proof-planning system. These enabled a degree of interactivity when running proof plans. The GUIs displayed proof plan trees, traces of the executions of CLAM proof methods at some default level of detail. Nodes in these trees were associated with method applications and were displayed as boxes labelled with method names. Edges in these trees were associated with goals: the parent edge of a method corresponded to the goal the method was applied to and the child edges of a method to any subgoals generated by the method. Goal formulas were not displayed by default but could be viewed in pop-up windows.

If more information was desired about a method application, a pop-up window could show a proof plan tree for the method, revealing the next greater level of detail. Alternatively, a method could be expanded in place into its next-level-down proof tree.

To help the user understand how planning was functioning, the user could check the status of method preconditions and method scores that the planner used to select methods.

Barnacle and XBarnacle were used and evaluated not only by researchers, but also by undergraduate students on a formal methods course. Users appreciated the graphical visualisation of proof trees and liked the ability to increase or reduce the level of detail, as the default level was often not the most useful [15].

The challenges of displaying tactic-and-subgoal proof trees have been considered in a number of provers. For example, PVS can generate two-dimensional display of proof trees, where the text of PVS tactics is shown, but goal formulas are only visible in pop-up windows that appear when goal symbols are clicked on. In my experience of using both Nuprl and PVS, I have found it most useful to view the full story, seeing both goals and tactics at once. I can see what is going on more quickly, and a full view can easily be printed and studied offline. The size of goal text usually forces a one-dimensional vertical layout of tree structure such as used in the proof-tree presentations in this chapter. Hopefully dynamic presentation techniques can help minimise the disadvantages of a one-dimensional layout.

4.7.2 The Orthogonal Hierarchies of Method Trees

In 2002, Bundy authored Blue Book Note 1411 with the title *Representing Orthogonal Hierarchies in Proof Plan Presentations*. The orthogonal hierarchies in question were the hierarchy of subgoals in the method-and-subgoal proof tree, and the hierarchy of methods and their constituent methods. He considered several alternative visual presentations of these two hierarchies. He remarked how the expansion and contraction of method applications in XBarnacle prevented one from seeing at a glance the relationship between a method application and its expansion. He advanced a preference for presenting higher-level methods and their constituent methods using nested boxes, and subgoals using edges between boxes with sequents labelling edges hidden by default. See Fig. 1.2 in Chap. 1 for an example of such a presentation.

Even with goals hidden, he remarked on the challenge of maintaining the readability of such presentations as the size of the tree increases. He noted obvious management techniques such as zooming in, making an internal node the root of the presentation and hiding certain subtrees. He observed that such techniques can be applied to both the method nesting hierarchy and the subgoal hierarchy.

4.7.3 IsaPlanner

Dixon and Fleuriot's IsaPlanner [8] was an exploration of importing proof-planning ideas into the Isabelle/Isar environment. IsaPlanner's *proof techniques*, enhanced versions of tactics, would output Isar declarative proof scripts when run on a proof goal. Particular techniques were responsible for generating script structure, and IsaPlanner provided support for unpacking a technique into lower-level constituent techniques. A graphical viewer was built for the generated proof plans along the lines Bundy had previously advocated (see Sect. 4.7.2), which used nested boxes to show how higher-level technique instances were composed of instances of more basic techniques.

4.7.4 Hiproofs and Proof Refactoring

Denney, Power and Tourlas [7] considered mathematical models of proof trees with hierarchies of both subgoals and methods as described above in Sect. 4.7.2. For brevity, they referred to them as *hiproofs*. Aspinall, Denny and Lüth [3] defined a simple grammar for hiproofs and a simple tactic language *Hitac* for generating hiproofs, and went on to present small-step and big-step operational semantics for Hitac. At the time, Aspinall had a strong interest in *proof re-engineering*, exploring ideas analogous to software re-engineering in the world of proofs: Whiteside, Aspinall, Dixon and Grov [24] defined a simple formal declarative proof script language reminiscent of Isabelle/Isar and gave it an operational semantics, building on the previous hiproof and Hitac work. They then considered a number of re-arrangements, *refactorings* of declarative proofs (e.g., turning a backward proof into a forward proof) and argued how these refactorings are formally correct.

Refactoring of software is used to improve its maintainability and understandability. Proof refactoring is of interest in this chapter because it could make proofs easier to understand.

4.7.5 HipCam and Tactician

Obua, Adams and Aspinall [18] produced two systems that can automatically generate hiproof versions of HOL Light proofs and then graphically display them. The issue is that the practice in HOL Light source files is to store the proof of each lemma as a maximally condensed single tactic. If our running example lemma were to be stored as an ML variable binding in the style used in HOL Light source files, it might look as shown in Fig. 4.9.

In a further paper [1], Adams explains how to use Tactician to refactor packed HOL Light proofs into sequences of individual tactic invocations on HOL Light's

```
let root_2_irrat_over_int = prove
  ⌜¬(∃m,n:ℤ. CoPrime(m,n) ∧ m * m = 2 * n * n)⌝
  ( D 0 THENM ExRepD
    THENM
    ( Assert ⌜2 | m⌝
      THENA (BLemma 'two_div_square' THENM Unfold 'divides' 0
             THENM AutoInstConcl []))
    THENM
    ( Assert ⌜2 | n⌝
      THENA (BLemma 'two_div_square' THENM All (Unfold 'divides')
             THENM ExRepD THENM Inst [⌜c * c⌝] 0
             THENM RWO "6" 4))
    THENM RWO "coprime_elim" 3 THENM FHyp 3 [5;6]
    THENM RWO "assoced_elim" 7 THENM D (-1) ...)
;;
```

Fig. 4.9 Single tactic proof

goal stack, with comments identifying the tree structure of the proofs. A HOL Light proof is usually initially produced by running a sequence of separate tactics that successively refine the top goal on a stack of remaining subgoals to prove. This refactoring simplifies stepped replay and viewing of HOL Light proofs, enabling novices to more easily study and learn from legacy HOL Light proofs and also helping with proof maintenance as revisions are made to HOL Light libraries. The refactoring can also be reversed, shortcutting the tedious process of transforming a stepped proof into a packed proof.

4.8 Technologies for Proof Presentation

As remarked in the introduction, the hope is that improved dynamic presentations of formal proofs will help to increase the ease with which formal proofs can be understood and will broaden the audience for formal proofs. There is the potential to engage those interested in learning topics that have been formalised and attracting the attention of those who initially are just casually interested. There is the potential too to support active users of theorem provers in rapidly coming up to speed on libraries in the systems they are using and learning too from libraries in other systems.

To achieve this, proof presentations must be accessible using standard universal technologies, i.e., web browsers. Also access must be fast; delays must be at most seconds. Could this level of performance be achieved by connecting to a web server running the relevant theorem prover? Would the server need to cache pre-processed presentation information?

It is desirable that presentations of proofs from a theorem prover be long-lasting, remaining accessible even after the theorem prover itself is no longer actively maintained, and perhaps after the point when running the prover on up-to-date hardware is problematic. This might steer the technology towards not relying on the theorem prover running and instead caching all relevant data. Simple hypertext presentations of Nuprl theories I developed 25 years ago are still readily browsable on the web, even though it is unlikely that the version of Nuprl I used when developing those theories still runs.

Hopefully some presentation technology could be shared across multiple theorem provers, to speed adapting it to new provers.

Presentation technologies would also need to address many of the issues not touched on here that are also highly desirable. For example, it should handle the pretty printing of formulas and terms, with control over often hidden information such as types, implicit arguments and implicit coercions, and the provision of hyperlinks or tool-tip hover-texts that explain pointed-to proof commands, definitions and lemma names. Modern programming IDEs, e.g., VSCode, provide good examples of how such features can be engineered. For example, if a programmer using VSCode wishes to see how a function being called is defined, they can easily instruct VSCode to insert a several-line scrollable buffer immediately below the function call position that displays the function's definition. Indeed, the preferred front end for the Lean prover uses VSCode, and a VSCode front end for Isabelle is being developed that might eventually replace the current jEdit front end.

4.9 Relationship Between Viewing and Editing Proofs

Whether or not fast dynamic presentation of theorem prover libraries uses a running instance of the theorem prover, it is certainly desirable that similar functionality be available to proof developers on the proofs they are currently working on. Good dynamic proof presentation should help the developer both focus on individual proof steps and keep a good awareness of the wider proof context. It should also help them more quickly understand why a proof step might not be running or checking as they expect, and so speed the completion of proofs.

As stressed at this chapter start, good dynamic proof presentation separates the concerns of how we input proofs, the required keystrokes and mouse clicks, from the concerns of how we view and understand proofs. This could lead to simpler, easier to learn, more robust proof guidance approaches than we currently have.

4.10 Further Related Work

The ACL2 theorem prover [13] has a number of options for controlling the kind of information and level of detail it shows in proofs. Theorems are proved using a single sophisticated automatic strategy. As this strategy runs, it prints subgoals with their tree addresses and between these gives natural-language descriptions of the reasoning techniques applied. When a proof fails, it also prints information on key steps in the failed proof that the user should first inspect in order to infer what guidance is missing. Perhaps a missing prior lemma is needed or perhaps the use of some existing previous lemma for rewriting needs to be disabled. Various options can reduce the amount of proof information printed or trace details of particular kinds of reasoning steps. Breakpoints can be set if one wants to interactively examine the prover state in particular parts of a proof attempt. To help the user appreciate how a proof is progressing, ACL2 can generate simultaneous alternate views summarising aspects of the evolving proof such as the subgoal tree structure or the applied rewrites.

A major difference between these dynamic presentation capabilities and those considered in this chapter has to do with the design purpose of the capabilities. With ACL2, the primary concern is with quickly figuring out why a proof fails and how to go about fixing it. In this chapter, a primary concern is for capabilities that help the user understand successful proofs. However, it is expected that capabilities that are good for this will also help interactive proof developers to track where they are in partial proofs and to debug faulty lines of reasoning.

Another difference concerns the extent to which the prover might construct some proof data structure that then separately can be traversed and inspected. The dynamic proof presentation discussed in this chapter assumes that such a data structure exists. With ACL2 the capabilities seem designed to largely avoid the construction of such data structures, perhaps because they would be prohibitively large for the formal verification applications ACL2 is typically used for. Interestingly, the developers of the Imandra theorem prover [19], which has automation strongly inspired by that of ACL2 and its predecessors, are experimenting with the benefits of creating hiproof-like proof data structures.

Siekmann et al. [21] describe a user interface for the ΩMEGA proof-planning system. ΩMEGA has a graph data structure for storing proofs that holds the multiple levels of detail of hierarchical proofs and additionally supports holding alternative proofs. Different kinds of edges in the graph record proof-tree subgoal hierarchy, how method applications are related to applications of their constituent methods, and how there might be multiple proofs of a given subgoal. In one panel, the interface presents a 2D layout of the interleaved subgoals and methods for a proof tree using different colours and shapes for nodes, but no visible method or goal information. Node colours and shapes distinguish whether, for example, a node represents a goal, a method or a primitive inference. Another panel shows a linearised natural-deduction view of the current proof, and when the proof is complete, a pop-up window can display a natural-language version of the proof. Some control is provided for restricting attention to parts of a proof. The alternate views in the different panels are hyperlinked so clicking at a point in one takes the user to the corresponding point in another.

Cairns and Gow [5] explored how students on a topology course handled semi-formal hierarchical proofs in the structured proof format advocated by Lamport [14]. This format is similar to formal declarative proofs in that justifications of higher-level steps are provided in lower-level proof blocks. Of particular interest to us is that the web presentation of the hierarchical proofs allowed viewers to selectively hide or expand the more-detailed proof levels. The responses from a preliminary survey of three students were mixed. The value of being able to control the level of detail was recognised, but the unfamiliarity of the format and an awkwardness of a numerical cross-referencing scheme were obstacles to the hierarchical proofs helping improve understanding of the proofs.

Wiedijk [25] describes the notion of a *formal proof sketch* that is derived from a formal declarative proof by omitting particular details in order to produce proofs that are easier to read. He illustrates this using formal proofs from the Mizar system. These formal sketches always preserve some essential formal structure of the corresponding full formal proofs.

Kalisyk and Wiedijk [12] describe the ProofWeb system that translates arbitrary procedural proofs in Coq into the declarative Fitch-style proofs as used in the Huth and Ryan textbook on formal verification [11]. Further, it enables users to develop incomplete proofs either by directly editing the Fitch-style proofs or by running Coq tactics on statements in the declarative proofs that have not yet been justified. Related later work by Wiedijk [27] presents a light-weight front end to HOL Light that runs in the Unix vi editor and that supports the creation of declarative proofs in the style of the Mizar prover (the main inspiration for the Isabelle/Isar declarative language). The user can mix typing the declarative text in full and just typing HOL Light tactics that run on unjustified steps and cause the system to extend the declarative proof.

Prover developers (e.g., for Isabelle, HOL4, Mizar, Coq, Lean, Metamath) do make efforts to have libraries browsable on the web, sometimes with useful hyperlinks for definitions and theorems. However, only in some cases are versions of libraries with proofs provided, and, when this happens, the proofs are usually just static proof scripts as recorded in proof script files. One exception is with the work by Tankink et al. [22, 23] on the Proviola system for Coq. This displays Coq source files in a web browser in such a way that clicking on a tactic step brings up a second pane displaying Coq's output from that step, typically a list of the subgoals generated. Another exception is with Pit-Claudel's recently released Alectryon tool [20], again for Coq libraries. As with Proviola, Coq's output can be viewed, but here the output is interleaved with the source, and users can click to unfold the display of further information or to fold the information currently displayed. Special comments can be added to source files to control what information about subgoals and subgoals parts is folded or unfolded by default.

4.11 Conclusions and Future Directions

This chapter has discussed how dynamic proof presentation could help ease understanding formal proofs and broaden the audience for formal proofs.

Some ideas for directions in which future research would be worthwhile are as follows.

Source Mark-Up for Presentation
Mark-up conventions are needed to indicate how blocks of proof are folded or unfolded by default, how comments might replace proof blocks, and how a proof has hierarchical structure that is not apparent from the proof syntax. Already provers such as Coq and Isabelle support mark-up for producing document versions of library files, and the Alectryon work [20] defines further mark-up for dynamic presentation options.

Exploring Further Proof Presentation Techniques
Once proof editors and proof viewers can be engineered to support dynamic proof presentation, there are opportunities for exploring ideas for presenting proofs beyond those discussed here, perhaps bringing in too the proof refactorings and transformations mentioned in Sects. 4.7.4, 4.7.5 and 4.10.

Handling Legacy Proofs
This is vital as many formal proofs have not been developed with the reader in mind, yet there is interest in understanding these proofs. Again, some combination of proof transformation technologies such as described in Sects. 4.7.4, 4.7.5 and 4.10, and dynamic presentation technologies is needed.

Proof Presentation Technologies
As discussed in Sect. 4.8, proof presentations should be viewable using web browsers and fast to access and navigate. How should this be engineered?

Prover Input Languages
Separating the demands of the languages for entering and displaying proofs opens up new opportunities for the input languages. In current proof languages, there are compromises between the different needs of ease of input, readability and suitability for instructing the theorem prover. With separation of demands, we can imagine simplified input languages suitable for novices and more sophisticated terse input languages for experts.

Exposing Proof-Tree Structure
Most interactive provers have tactics that transform a proof state consisting of a list or stack of unproven goals. In doing so, the tree-shaped hierarchical structure of proofs is obscured. It would be good if the presentation technology can expose this tree structure so it can help with proof understanding.

Handling Meta-Variables
While Nuprl tactics always refine a single unproven subgoal, tactics in other provers can simultaneously modify multiple subgoals in the unproven goal list. This can make creating tactic-and-subgoal tree presentations of proofs problematic. A prime example of when this happens is when the prover supports meta-variables—implicitly existentially quantified variables—in goals. Deep down in one branch of a proof, a tactic can instantiate a meta-variable that also occurs in other proof branches. How then should tactic-and-subgoal tree presentations make such non-local modifications of a proof tree evident?

Extracting Explanations from Tactic Runs
When a tactic encapsulates a significant amount of automation, it is desirable that the prover be able to explain tactic runs. If the tactic simply unpacks into calls of simpler tactics, then, as discussed in Sect. 4.5, showing a tactic-and-subgoal tree involving these simpler tactics could be appropriate. However, if it involves rewriting or involves calls of automated provers for first-order logic or arithmetic, then some kind of execution trace might be relevant. But such traces can often be far too detailed. What ways are there of structuring them so detail can be incrementally revealed?

Presenting Proof Terms
This chapter has not discussed the use of *proof terms* to describe proofs. This proof style is standard with the Agda proof assistant and popular with some Lean users. While proof terms precisely express the logical structure of proofs, they do so in a way less naturally familiar to most readers, and careful use of syntactic sugar and layout is needed to produce proofs with some of the readability of declarative proofs. How could a dynamic proof presentation approach make proof terms easier to understand?

As pointed out in Sect. 4.7, current group members Bundy, Aspinall and Fleuriot all have a significant amount of past experience in areas closely related to those discussed here. Further, for many years, Aspinall was the primary developer of the Proof General user interface for interactive theorem provers [2], and Fleuriot is a world-class expert in the Isabelle theorem prover and its Isar proof language. I hope this chapter will be a spur to some combination of us to now push forward on some of the topics listed here.

Acknowledgments I would like to thank the anonymous reviewers for their helpful recommendations.

References

1. Adams, M.: Refactoring proofs with Tactician. In: D. Bianculli, R. Calinescu, B. Rumpe (eds.) Software Engineering and Formal Methods - SEFM 2015 Collocated Workshops: ATSE, HOFM, MoKMaSD, and VERY*SCART, York, UK, September 7-8, 2015, Revised Selected Papers, *Lecture Notes in Computer Science*, vol. 9509, pp. 53–67. Springer (2015). URL https://doi.org/10.1007/978-3-662-49224-6_6
2. Aspinall, D.: Proof General: A generic tool for proof development. In: S. Graf, M.I. Schwartzbach (eds.) Tools and Algorithms for Construction and Analysis of Systems, 6th International Conference, TACAS 2000, Held as Part of the European Joint Conferences on the Theory and Practice of Software, ETAPS 2000, Berlin, Germany, March 25 - April 2, 2000, Proceedings, *Lecture Notes in Computer Science*, vol. 1785, pp. 38–42. Springer (2000). URL https://doi.org/10.1007/3-540-46419-0_3
3. Aspinall, D., Denney, E., Lüth, C.: Tactics for hierarchical proof. Mathematics in Computer Science **3**(3), 309–330 (2010). URL https://doi.org/10.1007/s11786-010-0025-6
4. Blanchette, J.C., Böhme, S., Paulson, L.C.: Extending Sledgehammer with SMT solvers. J. Autom. Reasoning **51**(1), 109–128 (2013). URL https://doi.org/10.1007/s10817-013-9278-5

5. Cairns, P.A., Gow, J.: A theoretical analysis of hierarchical proofs. In: A. Asperti, B. Buchberger, J.H. Davenport (eds.) Mathematical Knowledge Management, Second International Conference, MKM 2003, Bertinoro, Italy, February 16-18, 2003, Proceedings, *Lecture Notes in Computer Science*, vol. 2594, pp. 175–187. Springer (2003). URL https://doi.org/10.1007/3-540-36469-2_14
6. Constable, R.L., Allen, S.F., Bromley, H., Cleaveland, W., Cremer, J., Harper, R., Howe, D.J., Knoblock, T., Mendler, N., Panangaden, P., Sasaki, J.T., Smith, S.F.: Implementing Mathematics with the Nuprl Development System. Prentice Hall, NJ (1986). URL http://www.nuprl.org/book/
7. Denney, E., Power, J., Tourlas, K.: Hiproofs: A hierarchical notion of proof tree. Electr. Notes Theor. Comput. Sci. **155**, 341–359 (2006). URL https://doi.org/10.1016/j.entcs.2005.11.063
8. Dixon, L., Fleuriot, J.D.: A proof-centric approach to mathematical assistants. J. Applied Logic **4**(4), 505–532 (2006). URL https://doi.org/10.1016/j.jal.2005.10.007
9. Harrison, J.: Proof style. In: E. Giménez, C. Paulin-Mohring (eds.) Types for Proofs and Programs, International Workshop TYPES'96, Aussois, France, December 15-19, 1996, Selected Papers, *Lecture Notes in Computer Science*, vol. 1512, pp. 154–172. Springer (1996). URL https://doi.org/10.1007/BFb0097791
10. Holland-Minkley, A.M.: Planning proof content for communicating induction. In: Proceedings of the International Natural Language Generation Conference, Harriman, New York, USA, July 2002, pp. 167–172. Association for Computational Linguistics (2002). URL https://www.aclweb.org/anthology/W02-2122/
11. Huth, M., Ryan, M.: Logic in Computer Science: Modelling and Reasoning about Systems, 2 edn. Cambridge University Press (2004)
12. Kaliszyk, C., Wiedijk, F.: Merging procedural and declarative proof. In: S. Berardi, F. Damiani, U. de'Liguoro (eds.) Types for Proofs and Programs, International Conference, TYPES 2008, Torino, Italy, March 26-29, 2008, Revised Selected Papers, *Lecture Notes in Computer Science*, vol. 5497, pp. 203–219. Springer (2008). URL https://doi.org/10.1007/978-3-642-02444-3_13
13. Kaufmann, M., Manolios, P., Moore, J.S.: Computer-Aided Reasoning: An Approach. Kluwer Academic Publishers (2000)
14. Lamport, L.: How to write a proof. Tech. Rep. 94, DEC Systems Research Center (1993). URL https://www.hpl.hp.com/techreports/Compaq-DEC/SRC-RR-94.pdf
15. Lowe, H., Bundy, A., McLean, D.: The use of proof planning for co-operative theorem proving. J. Symb. Comput. **25**(2), 239–261 (1998). URL https://doi.org/10.1006/jsco.1997.0174
16. Lowe, H., Duncan, D.: XBarnacle: Making theorem provers more accessible. In: W. McCune (ed.) Automated Deduction - CADE-14, 14th International Conference on Automated Deduction, Townsville, North Queensland, Australia, July 13-17, 1997, Proceedings, *Lecture Notes in Computer Science*, vol. 1249, pp. 404–407. Springer (1997). URL https://doi.org/10.1007/3-540-63104-6_39
17. de Moura, L.M., Kong, S., Avigad, J., van Doorn, F., von Raumer, J.: The Lean theorem prover (system description). In: A.P. Felty, A. Middeldorp (eds.) Automated Deduction - CADE-25 - 25th International Conference on Automated Deduction, Berlin, Germany, August 1-7, 2015, Proceedings, *Lecture Notes in Computer Science*, vol. 9195, pp. 378–388. Springer (2015). URL https://doi.org/10.1007/978-3-319-21401-6_26
18. Obua, S., Adams, M., Aspinall, D.: Capturing Hiproofs in HOL Light. In: J. Carette, D. Aspinall, C. Lange, P. Sojka, W. Windsteiger (eds.) Intelligent Computer Mathematics - MKM, Calculemus, DML, and Systems and Projects 2013, Held as Part of CICM 2013, Bath, UK, July 8-12, 2013. Proceedings, *Lecture Notes in Computer Science*, vol. 7961, pp. 184–199. Springer (2013). URL https://doi.org/10.1007/978-3-642-39320-4_12
19. Passmore, G.O., Cruanes, S., Ignatovich, D., Aitken, D., Bray, M., Kagan, E., Kanishev, K., Maclean, E., Mometto, N.: The Imandra automated reasoning system (system description). In: N. Peltier, V. Sofronie-Stokkermans (eds.) Automated Reasoning - 10th International Joint Conference, IJCAR 2020, Paris, France, July 1-4, 2020, Proceedings, Part II, *Lecture Notes in Computer Science*, vol. 12167, pp. 464–471. Springer (2020). URL https://doi.org/10.1007/978-3-030-51054-1_30

20. Pit-Claudel, C.: Untangling mechanized proofs. In: R. Lämmel, L. Tratt, J. de Lara (eds.) Proceedings of the 13th ACM SIGPLAN International Conference on Software Language Engineering, SLE 2020, Virtual Event, USA, November 16-17, 2020, pp. 155–174. ACM (2020). URL https://doi.org/10.1145/3426425.3426940
21. Siekmann, J.H., Hess, S.M., Benzmüller, C., Cheikhrouhou, L., Fiedler, A., Horacek, H., Kohlhase, M., Konrad, K., Meier, A., Melis, E., Pollet, M., Sorge, V.: LΩUI: Lovely ΩMEGA User Interface. Formal Asp. Comput. **11**(3), 326–342 (1999). URL https://doi.org/10.1007/s001650050053
22. Tankink, C., Geuvers, H., McKinna, J., Wiedijk, F.: Proviola: A tool for proof re-animation. In: S. Autexier, J. Calmet, D. Delahaye, P.D.F. Ion, L. Rideau, R. Rioboo, A.P. Sexton (eds.) Intelligent Computer Mathematics, 10th International Conference, AISC 2010, 17th Symposium, Calculemus 2010, and 9th International Conference, MKM 2010, Paris, France, July 5-10, 2010. Proceedings, *Lecture Notes in Computer Science*, vol. 6167, pp. 440–454. Springer (2010). URL https://doi.org/10.1007/978-3-642-14128-7_37
23. Tankink, C., McKinna, J.: Dynamic proof pages. In: C. Lange, J. Urban (eds.) Proceedings of the ITP 2011 Workshop on Mathematical Wikis, Nijmegen, The Netherlands, August 27th, 2011, *CEUR Workshop Proceedings*, vol. 767, pp. 45–48. CEUR-WS.org (2011). URL http://ceur-ws.org/Vol-767/paper-08.pdf
24. Whiteside, I., Aspinall, D., Dixon, L., Grov, G.: Towards formal proof script refactoring. In: J.H. Davenport, W.M. Farmer, J. Urban, F. Rabe (eds.) Intelligent Computer Mathematics - 18th Symposium, Calculemus 2011, and 10th International Conference, MKM 2011, Bertinoro, Italy, July 18-23, 2011. Proceedings, *Lecture Notes in Computer Science*, vol. 6824, pp. 260–275. Springer (2011). URL https://doi.org/10.1007/978-3-642-22673-1_18
25. Wiedijk, F.: Formal proof sketches. In: S. Berardi, M. Coppo, F. Damiani (eds.) Types for Proofs and Programs, International Workshop, TYPES 2003, Torino, Italy, April 30 - May 4, 2003, Revised Selected Papers, *Lecture Notes in Computer Science*, vol. 3085, pp. 378–393. Springer (2003). URL https://doi.org/10.1007/978-3-540-24849-1_24
26. Wiedijk, F. (ed.): The Seventeen Provers of the World, Foreword by Dana S. Scott, *Lecture Notes in Computer Science*, vol. 3600. Springer (2006). URL https://doi.org/10.1007/11542384
27. Wiedijk, F.: A synthesis of the procedural and declarative styles of interactive theorem proving. Logical Methods in Computer Science **8**(1) (2012). URL https://doi.org/10.2168/LMCS-8(1:30)2012

Chapter 5
Proof Mechanization: From Dream to Reality

Jacques D. Fleuriot

Abstract Two research strands, namely proof planning and geometric reasoning, from my early days in the DREAM Group have influenced my thinking and work over the years. I explore some of the motivations, central ideas and achievements attached to these. Along the way, I try to weave a unifying thread about building tools and approaches that help us explore and mechanize proofs, while reminiscing about some of the events surrounding the research.

5.1 Prologue

In 1997, Larry Paulson, my PhD supervisor at Cambridge, suggested that I should give a seminar in the DREAM Group on the mechanization of proofs from Newton's *Principia Mathematica* [13]. He felt that Alan Bundy's group was the perfect place for me to give my first talk about the work to the outside world. Little did I know that 2 years or so later, this would become my new home.

I was a bit nervous at the prospect of travelling all the way to Edinburgh but was reassured by the fact that Richard Boulton, who finished his PhD under Mike Gordon a few months after I moved to Cambridge, was working in the DREAM Group on a project that involved linking Edinburgh's *Clam* proof planner with Cambridge's HOL proof assistant [2]. He kindly offered to meet me at the train station and take me to the department. Alan was an excellent host on the day and showed genuine interest in what I was doing, which as a PhD student was a great confidence booster. The day went really well and I received some fantastic feedback, although I still remember a probing question from Alan Smaill about my somewhat loose use of "constructive" when talking about some of my geometric proofs.

In hindsight, the only thing I should have been nervous when it came to the whole trip was my decision to travel to Edinburgh and back in 1 day, as I almost missed

J. D. Fleuriot (✉)
Artificial Intelligence and Its Applications Institute (AIAI), School of Informatics, University of Edinburgh, Edinburgh, UK
e-mail: jdf@ed.ac.uk

G. Michaelson (eds.), *Mathematical Reasoning: The History and Impact of the DReaM Group*, https://doi.org/10.1007/978-3-030-77879-8_5

the last connecting train in Peterborough on the wasy back that evening. Over the years, despite ups and downs, I could not have wished for a better environment than the DREAM Group when it comes to exploring new ideas, regardless of how ϵ-baked they might be (for $1 \geq \epsilon \geq 0$, as per the Blue Book Notes[1]). In what follows, I revisit some of these ideas that, in some form or another, have been shaped by my fellow DREAMers and examine how we mechanized some of our mathematical dreams.

5.2 Proof Planning

I must admit from the outset that I had no familiarity with proof planning before coming to Edinburgh. However, as most DREAM Group members were working in the area, my induction (pun intended) was fairly painless and, coming from the interactive world of Isabelle [36], it was a bit of an eye-opener how challenging the automation of inductive proof could be.

5.2.1 *Nonstandard Analysis in λClam*

Fairly quickly, it seemed to me that some of the ideas from proof planning could probably be applied to the automation of proofs in nonstandard analysis (NSA) [39], where the ϵ–δ formulation of concepts such as sequences, limits and derivatives is replaced by more intuitive ones involving infinitesimals, infinite numbers and an *infinitely close* relation. The latter, an equivalence relation that holds when two numbers only differ by an infinitesimal, made many of the mechanized proofs more "calculational" and resulted in distinct reasoning patterns (that I had noticed when working interactively in Isabelle). All this hinted at the possibility of using rewriting techniques such as rippling [4]. Moreover, I had already mechanized in Isabelle some classic results from (standard) real analysis such as Rolle's Theorem and the Intermediate Value Theorem (IVT), and it seemed to me that their proofs could be recast into ones that would involve recursive approximations whose properties could be proved inductively via proof planning. Then, as the recursion was extended to an infinitely large number of steps, the approximations would be shown to be infinitely close to the expected results by algebraic reasoning.

This was all very exciting to me and, as some further good news, I had a new PhD student, Ewen Maclean who was keen to work on the topic. His undergraduate degree in mathematics and his freshly minted MSc in AI from Informatics, where he had been exposed to proof planning and other automated reasoning concepts, made him the perfect match. We decided the work would be done in *λClam* [37] as it

[1] https://dream.inf.ed.ac.uk/computing/blue_notes.html.

supported higher-order proof plans. Ewen's first important result, which showed that we had a viable approach, was to fully proof plan the IVT [21]. This was eventually followed by other non-trivial results such as Rolle's Theorem and the Mean Value Theorem, all of which could be captured by a collection of general plan specifications. Along the way, he figured out ways to deal with tricky aspects such as transferring results between the standard and nonstandard domains, also via general proof plans.

Looking back, the nonstandard analysis proofs [22] that Ewen Maclean managed to get through λ*Clam* are a tour-de-force that deserves wider recognition. Aside from building a comprehensive proof planning framework for nonstandard analysis that incorporated techniques such as coloured rippling and critics, working with λ*Clam* tested his mettle throughout the PhD due to the brittle and buggy nature of λProlog. Much time was spent on understanding and then fixing bugs that were due to the experimental nature of λProlog or, when nothing else worked, waiting for a fix from its developers. We often worried that we were building a whole infrastructure on unsteady grounds that could bring his research to a halt.

The experience with λ*Clam* convinced me that a new proof planning framework was worth pursuing and that Isabelle (and more specifically its higher-order object logic) would be the right setting for this.[2] This lead to my work with Lucas Dixon on ISAPLANNER [10], which was funded by my Fast Stream grant.[3]

5.2.2 ISAPLANNER

The general philosophy behind ISAPLANNER was that it would be a new proof planning system that would benefit both from the rigorous logical framework of Isabelle and from PolyML [23], the robust programming language underpinning it. One of my main motivations was to create a system that could plan but also execute the proof plans within Isabelle itself to produce object-level proofs (something that λ*Clam* could not do as it did not have an associated theorem prover). Moreover, it should also be easy to add new tools, methods and critics to the system. This led to the development of (what we called) a technique language. It resembled Isabelle's tactic language to some extent while being tailored to proof planning. However, aside from the high-level programmability, since Isabelle exposed an API to its proof tools and underlying data structures, the user would also be allowed to create more advanced techniques with the help of PolyML. These would have access to Isabelle's term and theorem structures, could hook into its simplifier at

[2]It is worth noting here that I was the only Isabelle user in the DREAM Group at the time. Yet, there still was encouraging support when I spoke about my plans.

[3]EPSRC's Fast Stream scheme was for new lecturers and could provide just about enough to fund a PhD studentship for 3 years. In this case, this was around £60k and the original idea had been to link λ*Clam* to Isabelle.

specific points, and use the underlying search algorithms among a plethora of other possibilities.

Despite the lofty goals, the development of the system proceeded swiftly through a focused process involving Lucas and I, and also the constructive feedback we would get from the DREAMers at regular intervals. The DREAM Group, without any doubt, was the best place to start a new proof planner from scratch. There were multiple challenges along the way, including issues such as being unable to modify the simplification machinery of Isabelle to suit our needs due to its inherent complexity. However, such setbacks motivated us to investigate alternative ways of achieving our goals and, in the end, aside from developing ISAPLANNER, we also ended up enhancing Isabelle. A few of the highlights, aside from a nice PhD thesis [9], that were achieved through the Fast Stream project included:

- Higher-order rippling [11], which could deal with a case study in ordinal arithmetic setup specifically for it by Dennis and Smaill [8]. Moreover, the process was much faster, e.g., an exponentiation theorem that took over 5 min $\lambda Clam$ took 2 s in IsaPlanner to be fully planned and output as an Isabelle proof script.
- An inductive theorem prover that, aside from rippling, included lemma conjecturing and subterm generalisation.
- The generation of intelligible proof scripts in the Isabelle/Isar language [48], thereby presenting proofs to the user in a comprehensible fashion. Importantly, this also aligned with our "manifesto" [12] for a proof-centric approach to proof assistants, which asked for a departure from the goal-state-centric view and motivated some of Wenzel's work on achieving asynchronous proof editing in Isabelle [47].
- A generic ordered rewriting system for Isabelle that could be accessed via Isar. The machinery could be parameterised by a user-defined set of rules and user-written orderings for rule application.
- An improved equational substitution tactic for Isabelle, which went beyond what was possible using the simplifier (or the existing substitution tactic) by allowing the introduction and instantiation of meta-variables. This resulted from the need to support middle-out reasoning [16] and proof critics in ISAPLANNER. This tactic is still the main one in use in Isabelle to this day.

As an additional remark about enhancing inductive proof automation in interactive theorem provers, the framework [54] built by Sean Wilson in Coq [1], though not as well known as ISAPLANNER, is also worth mentioning. This work was very much inspired by proof planning in the way it identified common proof patterns and then looked into how they could be automated in a dependent types setting. Although the motivation was not to build a full proof planner in Coq, notions such as dynamic rippling and generalisation were implemented and shown to be effective in discharging non-trivial proof obligations. More details can be found in Sean Wilson's PhD thesis [52].

5.2.2.1 Interlude: Graduation Time for IsaPlanner

In early Spring 2004, some discussion started in the DReam Group about whether we should consider abandoning λ*Clam* and move to IsaPlanner due to numerous issues with λProlog. There was a growing feeling that we were at an impasse with the system with Louise Dennis, for instance, stating that "λ*Clam* is a trial not a joy to work with and the day-to-day legwork is unrewarding both in terms of user satisfaction and research output ...".[4]

The potential switch was discussed at the annual review meeting of the DReam Group's Platform Grant, and the recommendation from its advisory board was to make IsaPlanner the main vehicle for proof planning research in the DReam Group. Following the meeting, Alan Bundy asked me whether I was happy for this to happen. In some ways, this meant surrendering control of IsaPlanner's future direction, but I was fine with this as I felt the system was ready for broader use and the injection of new ideas that would take it to the next step. Looking back, I am extremely proud that my small grant and collaboration with Lucas Dixon produced a system that became part of the fabric of the DReam Group's research and allowed DReaMers, both seasoned and new PhD students to explore new research topics.

5.3 Geometric Reasoning: Marrying Discovery and Proof

Another strand I will now discuss involves our work on geometric reasoning. This is longstanding and, over the years, has focused on aspects ranging from axiomatic investigations to formal verification and enabled members of the DReam Group working in the area to be at the forefront of the Automated Deduction in Geometry (ADG) community. In the next few sections, while I look back on theorem-proving work, I will also focus on geometric discovery as an important tool for exploration and mechanization.

5.3.1 Of Chairs, Tables and Beer Mugs: Hilbert's Axiomatics

Hilbert's *Grundlagen der Geometrie* (Foundations of Geometry) [17] is widely considered to be of substantial historical interest in mathematics. Although Dehlinger et al. [7] were the first to examine its constructive aspects in Coq [1], our mechanical investigation is probably the deepest and longest running one and, over the years, it has uncovered a variety of hitherto unknown aspects.

Our research was initially motivated by the widely held view that the *Grundlagen* is highly rigorous, despite its heavy reliance on prose-based definitions and proofs.

[4]Email communication to the investigators on the Platform Grant.

Hilbert's claim that points, lines and planes could be replaced by tables, chairs and beer mugs using his axiomatic approach was generally taken at face value, leading prominent mathematicians like Weyl to say that Hilbert's deductions had no gaps [49] and, like Hilbert himself had asserted, that there was no need for any geometric intuition.

The reality, starting with the joint work with Laura Meikle as part of her final-year undergraduate project, was shown to be somewhat different [27]. The mechanization of Hilbert's axiomatics in Isabelle uncovered that there were difficult to explain gaps and that Hilbert did make implicit geometric assumptions in some of the proofs when he accompanied these with diagrams. For instance, his simply stated Theorem 3 about the existence of a collinear point between any two distinct points relied on the distinctness of a constructed point that was evident from an accompanying diagram but was elided from the proof despite being crucial (as was clearly demonstrated by our mechanization). This implicit use of diagrams to guide the reader through the reasoning provided evidence that contrary to popular belief Hilbert's proofs did rely on geometric intuition at times.

As we carried out more proofs, it became fairly clear that many aspects relating to the distinctness of points and lines, incidence, collinearity and planarity were often left implicit in Hilbert's reasoning. Although some of these omissions (e.g., the one mentioned above) were difficult to explain, in later work with Phil Scott, we managed to come up with an ad-hoc set of lemmas that could be used to manually justify a large number of gaps related to distinctness in a fairly general fashion [40]. This strongly indicated that a systematic and automatic way of carrying out Hilbert's implicit reasoning might be possible. With this in mind, we decided to see whether we could automatically plug the gaps in his proof and mechanize these as faithfully as possible to the Grundlagen. This led us eventually to (revisit and) incorporate geometric discovery ideas into our theorem-proving process to tackle Hilbert's work (see Sect. 5.3.2.2). But, before we go there, we first revisit our work on Geometry Explorer, as this has some bearing on how we decided on the approach.

5.3.2 Geometric Discovery

Our work on geometric discovery started by looking at fully automatic theorem proving in a way that combined dynamic geometry, automated theorem proving and discovery, and diagrammatic proofs.

5.3.2.1 Geometry Explorer

Geometry Explorer [53] was created as part of Sean Wilson's final-year project. In this work, a fully automatic theorem prover based on the full-angle method by Chou et al. [6] was rationally reconstructed and extended to deal with dynamic geometric reasoning and diagrammatic proofs in Euclidean geometry. The prover consisted of a Prolog-based engine that could do backward-chaining from the current goal using rules such as "$\angle[AB, CD] = \angle[AB, EF]$ if $CD \parallel EF$", where the full angle $\angle[AB, CD]$ can be thought of intuitively as the rotation required to make the line AB parallel to line CD. However, since backward-chaining alone was not powerful enough to find a proof for non-trivial theorems, the system also incorporated forward-chaining. The latter was applied using rules such as "If A, B and C are collinear then $AB \parallel BC$" to all the known geometry facts in the system's database, known as a Geometry Information Basis (GIB), to discover other geometric facts. These new facts would then be inserted into the GIB and forward-chaining applied again. This process happened repeatedly until no new facts could be found, with the hope that the augmented GIB would be enough to find a proof using backward-chaining. Geometry Explorer, despite being developed within the constraints of an undergraduate project, was competitive with the original system by Chou et al. and was able to prove about 100 of 110 benchmarks theorems [5] featuring many from the American Mathematical Monthly and the International Mathematical Olympiad (note that the 10 unproven theorems were out of scope for the system because they seemed to use rules, e.g., about the orthocentres of triangles, that were not part of the full-angle method). This combination of discovery and proof ended up influencing our later work on integrating the two within an interactive theorem-proving context (see next section). Next, we illustrate a few aspects of the approach in Geometry Explorer using the following example:

The Nine-Point Circle Theorem (NPCT). Let the midpoints of the sides AB, BC, and CA of $\triangle ABC$ be E, F, and G, respectively, and AD be the altitude on BC. Show that D, E, F, and G are cyclic.

Initially, the input and output of Geometry Explorer were textual and given at the command line. We had designed a small input language—in effect a domain-specific language (DSL)—to state the problems at a level of abstraction suitable for the full angle method, thereby avoiding the need to specify the problem directly in Prolog (see Fig. 5.1). The output, when a proof was found, was in the form of a step-by-step justification showing each of the rules that was used and could also be exported as a LaTeX proof. Although this way of working with the system seemed more adequate than for a fully automatic theorem prover, we felt that a more visual way of interacting with the system was worth exploring. I was especially keen to see whether we could incorporate notions from dynamic geometry [19], after seeing several impressive demos of Cinderella [38] at Automated Deduction in Geometry conferences.

Our aim thus became the creation of a dynamic geometry interface that would allow the user to construct diagrams fully visually using ruler and compass tools in the GUI and then to use this to explore both the proof and discovery process. Unlike pen-and-paper drawings though, the constructions would be manipulable and allow the user to move points around, with the system ensuring that constraints imposed by the construction (e.g., collinearity, perpendicularity, etc.) were maintained. The position of dependent constructions would then update automatically, generating different diagrammatic instances for the same geometric statement (see Fig. 5.1).

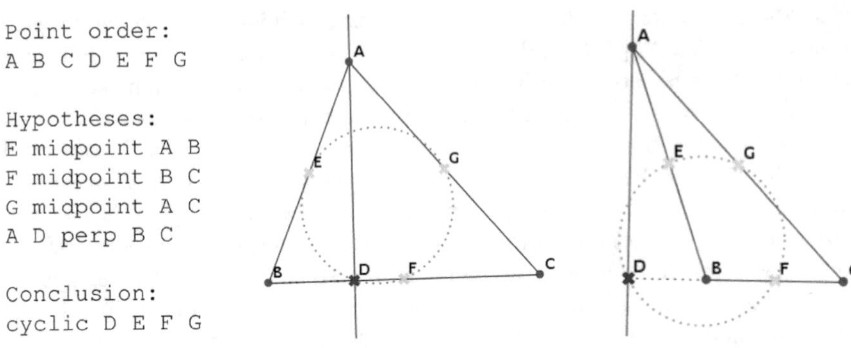

Fig. 5.1 Left: Specifying the Nine-Point Circle Theorem in the geometric DSL, with the point order referring to when the points are introduced during the construction. Middle and Right: Screenshots of alternative configurations created by manipulating the diagrammatic construction. Notice that in the middle F is between D and C, while on the right D has been moved off the line segment BC and the conjectured circle is no longer contained within $\triangle ABC$

This approach, once realised, allowed the exploration of diagrammatic properties and helped users discover new conjectures in ways that are not possible with static diagrams. After specifying a conjecture diagrammatically, invoking the integrated full-angle method theorem prover triggered a search for a proof via forward- and backward-chaining.

Although finding a proof was the main goal, discovering interesting facts about a conjecture was often the most exciting aspect since some of these could be quite surprising. Visualisations, built using Graphviz [20], enabled the user to look at the chain of discoveries, represented either textually or visually (as shown in Fig. 5.2) as nodes in a graph. The ability to explore discovered facts was missing from the previous work on full angles (an aspect that had made the results of the prover hard to debug during the development phase) and, as far as we could tell, our graphical presentation was also novel. As a side note here, we remark that Sean Wilson's experience with graph layout would come in handy several years later when we worked together on building the WorkflowFM diagrammatic process composer [35].

Looking back, although the research was quite successful and had much scope for future work, that we did not carry on with the development of Geometry Explorer seems a bit of a missed opportunity. Having said this, some of the insights I gained when it comes to marrying discovery and proof became extremely valuable when dealing with implicit aspects in Hilbert's proofs, which I discuss next.

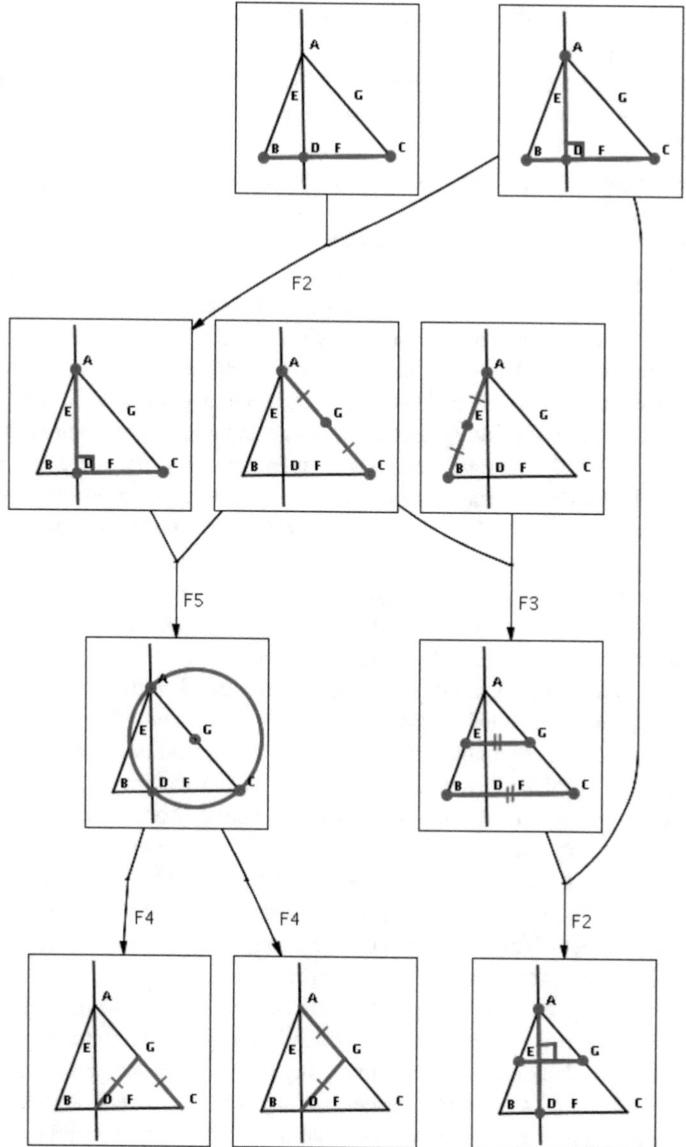

Fig. 5.2 Discovering facts through forward-chaining from the hypotheses of the NPCT

5.3.2.2 Hilbert's Implicit Reasoning and Idle-Time Proof Discovery

As mentioned in Sect. 5.3.1, mechanizing Hilbert's incidence reasoning was problematic. This motivated us to tackle it head-on when working on the proofs in HOL Light [15].[5]

At first sight, the problem looked fairly distinct from the fully automatic one explored with Geometry Explorer. In HOL Light, the mechanization used a declarative (or structured) proof approach based on the Mizar Light declarative language [51]. The user states the formulas that connect premises to the desired conclusion and justifies these by invoking the appropriate tactics interactively in the proof assistant. This results in proofs whose logical structure can be analysed and compared with Hilbert's pen-and-paper prose. However, while this proof style provides a readable version of Hilbert's text, it seemed to leave little scope for automatic, "unsupervised" exploration since the user usually works out the sequence of inference by hand. The challenge was to integrate proof discovery as a tool that would complement rather than interfere with the user's theorem-proving workflow.

Our approach thus involved a non-disruptive route, whereby a discovery tool would work in the background to derive facts that followed from the current context and make them available for use in proof as the user saw fit [43]. In order to do this, as is often the case in automated reasoning, we first had to come up with more appropriate representations for incidence reasoning.

Hilbert's first group of axioms (Group I), concerned with incidence relations between the primitives called points, lines and planes, requires 10 of his 23 axioms (as obtained after splitting conjunctions). Thus, one would expect these to feature significantly in proofs. However, although this is the case with the mechanized proofs, they are seldom mentioned in Hilbert's actual proofs. Moreover, Pasch's Axiom (from Group II), which asserts that any line that enters a triangle ABC on one side and does not meet any of the vertices must leave by one of the other two sides (see Fig. 5.3), was one of the important general incidence rules often used implicitly by Hilbert. However, when mechanizing Hilbert's reasoning, much work involved discharging the preconditions of this axiom.

Our refactoring of Pasch's Axiom and the axioms of Group I involved a reformulation in terms of the collinearity and planarity of sets of points rather than Hilbert's primitive incidence relations involving points, lines and planes (denoted by `on_line` and `on_plane` in the mechanization). Thus, new versions were derived that aimed to capture the combinatorial nature of incidence reasoning and open up the possibility of using set operations to facilitate the automated discovery of relevant geometric facts. In particular, by defining the collinearity and planarity of sets of points as

[5] Although our previous work on the Grundlagen had been in Isabelle, by 2009 I had decided to move some of our research to HOL Light, partially because of the flexibility of the system when it comes to the rapid development of reasoning tools.

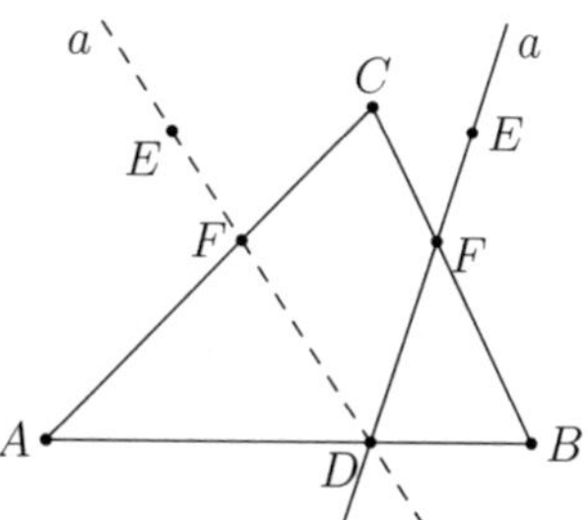

Fig. 5.3 Pasch's Axiom

$$\texttt{collinear}\ Ps \equiv \exists a.\forall P.P \in Ps \implies \texttt{on\_line}\ P\ a$$

and

$$\texttt{planar}\ Ps \equiv \exists a.\forall P.P \in Ps \implies \texttt{on\_plane}\ P\ a$$

the following version of Pasch's Axiom was obtained:

$$\begin{aligned}&\neg\texttt{collinear}\,\{A, B, C\} \wedge \neg\texttt{collinear}\,\{A, D, E\} \wedge \neg\texttt{collinear}\,\{C, D, E\}\\ &\wedge \texttt{planar}\,\{A, B, C, D, E\} \wedge \texttt{between}\ A\ D\ B\\ &\implies \exists F.\ \texttt{collinear}\,\{D, E, F\} \wedge (\texttt{between}\ A\ F\ C \vee \texttt{between}\ B\ F\ C)\end{aligned}$$

and new Group I incidence theorems derived, such as

$$S \subseteq T \wedge \texttt{collinear}\ T \implies \texttt{collinear}\ S$$

$$P \in S \wedge P \in T \wedge \texttt{collinear}\ S \wedge \texttt{collinear}\ T \implies \texttt{planar}\ (S \cup T)$$

With such incidence rules set up, these could now be used as parameters to forward-chaining discovery engines as described briefly next.

Inspired by the approach used in Geometry Explorer, Phil Scott and I decided to develop forward-chaining algorithms that would exhaustively search for (i.e., discover) new facts—including, for instance, the hoped-for assumptions that would enable the application of Pasch's Axiom to happen automatically—and integrate these into the proof in an implicit way, thereby relieving the user of the burden of discharging them. The proof script, we also hoped, would then match Hilbert's proof more closely. This would happen by exploiting what we called *idle time* during interactive proof [43].

Generally, in proof assistants, automation needs to be explicitly invoked by the user during proof. However, our experience showed that interactive proof development involves much thinking and elaboration time, as well as writing down the statements and proof commands. During that time, the proof engine is mostly idle, making it available to run other automated tools concurrently. As we wanted any such tool to complement the user's interactive and declarative development of

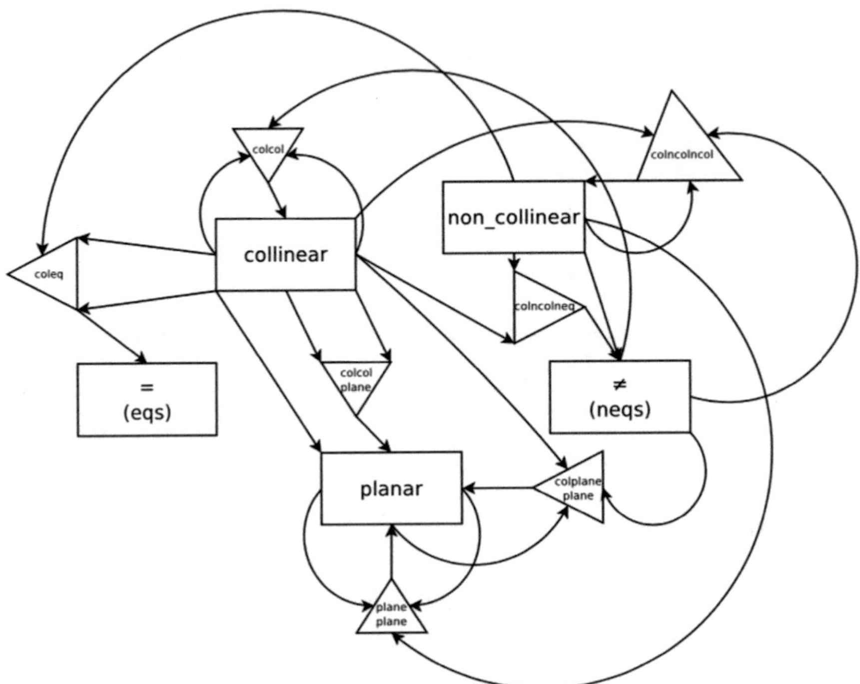

Fig. 5.4 Background incidence discoverer with rectangles representing five classes of derived data and triangles representing inference rules

the proof (i.e., not disrupt their workflow), we designed the automation to use the current proof context and derive facts that might interest the user, or even solve the goal outright, while they investigate their own chains of deduction independently.

In the case of Hilbert's geometry, based on our understanding of his proofs, we worked out that incidence reasoning is naturally partitioned into five classes of data with incidence rules connecting how they could be derived from each other as shown in Fig. 5.4. So, for instance, rule `colncolncol` inferred new non-collinear triples from a collinear set and another non-collinear triple:

$$\texttt{collinear}\ S \wedge \neg\texttt{collinear}\ \{A, B, C\} \wedge X, Y, A, B \in S \wedge X \neq Y$$
$$\Longrightarrow \neg\texttt{collinear}\ \{C, X, Y\}$$

As an example, given a context with facts $A \neq C$, $D \neq E$, $\texttt{collinear}\{A, B, C, D, E\}$ and $\neg\texttt{collinear}\{A, B, P\}$, triangles ACP and DEP can be discovered since $\neg\texttt{collinear}\{A, C, P\}$ and $\neg\texttt{collinear}\{D, E, P\}$ follow using the rule.

In order to capture these patterns of forward reasoning, a generic algebraic language was specifically designed for discovery [44]. By taking a general approach, incidence automation, as illustrated in Fig. 5.4, became just one possible (instance of a) handcrafted discoverer, similar to how a technique language was devised in IsaPlanner on top of Isabelle's tactic language. This led to a framework consisting of discovery engines that could be composed [42] and output facts to a database, whose contents could be applied by the user via new declarative language primitives such as `obviously` and `clearly` [41] to highlight their implicitness.

With this setup, we managed to tackle most of the gaps in Hilbert's reasoning. Interestingly, the incidence discoverer found exciting ways to apply Pasch's axiom, including a novel, alternative proof for Hilbert's fourth Theorem (which states that there is a point that lies between two other distinct ones on a line) [43]. In the end, the discoverer reduced the number of formalised proof steps by around a factor of 10 and enabled declarative proofs whose steps matched Hilbert's prose ones very closely, leading to a de Bruijn factor [50] of almost one.[6] Extensive details of the discovery language, iterative generations of facts and the integration of discovery engines among other aspects can be found in Phil Scott's PhD thesis [41], to which I refer the interested reader.

5.3.3 Computational Geometry

As a final part of the current tour, I briefly examine some work on mechanizing geometric algorithms, which combine formal verification and formalized mathematics. As previously discussed, initial work with Laura Meikle on mechanising Hilbert's axiomatics in Isabelle had shown that his proofs did rely on intuition. This led us to believe that the confidence in the correctness of computational geometry (CG) algorithms, which is often argued semi-formally with the use of diagrams, could be suspect. Little formal verification work had been done at the time in the field, so we decided to start our investigation by formalising the well-known Graham Scan (GS) algorithm. This is a straightforward-looking algorithm that computes the convex hull of a set of points in two-dimensions [14], i.e., it finds the smallest convex polygon C such that for a set of points P, every point in P either lies inside C or on its boundary.

As we wished to investigate the algorithm in its usual imperative form—as opposed to a functional programming style, which is relatively common when working in Isabelle/HOL [30]—we decided to base it on an existing mechanization of Floyd–Hoare Logic [31], which would allow us to capture its usual pseudo-code presentation fairly faithfully [28]:

[6] In a nutshell, the de Bruijn factor is the ratio of the size of a formalization of a mathematical text to the size of its pen-and-paper original.

```
{ordered P ∧ 3 ≤ length P ∧ distinct P ∧ ¬all_collinear P}
     i := 0;
     C := [hd P, last P];
     WHILE i < length P
     INV {Loop Invariant}
     DO
       IF Left_turn C$_1$  C$_0$  P$_i$
       THEN C := P$_i$  # C;
            i := i+1
       ELSE C := tail C
       FI
     OD
{(butlast C) isConvexHull P}
```

Although the above algorithm looks simple—its only non-trivial operation being the test `Left_turn` A B C that, for any three points A, B and C, is true if C lies to the left of the directed line from A to B—its verification in Isabelle was far from straightforward. One of the hardest parts, as anyone familiar with formal verification using Hoare Logic might attest, was figuring out the loop invariant. Even with the help of a pen-and-paper proof by O'Rourke [32], the formal verification proved very demanding, as only 5 components of the loop invariant followed and another 11 had to be worked out in order to enable a fully formal proof, including, for instance:

$$l < \texttt{length}\, C - 1 \wedge j < k \wedge k < l \Longrightarrow \texttt{Left\_turn}\, C_l\, C_k\, C_j$$

which states that if we travel along say C_l to C_k, then we must make a left turn with respect to vertices of the hull added after C_k. Various verification conditions were then derived automatically using the Hoare Logic machinery of Isabelle. The one about the body of the program preserving the loop invariant as long as the condition of the WHILE loop held was the hardest to discharge. This is because it involved a case-split about whether there was a left-turn or not when constructing the hull one point at a time. At the time, given how painful the process was, Laura Meikle and I wondered whether the work by Stark and Ireland on the automatic discovery of loop invariants [45] might provide a way of helping such convoluted geometric assertions. Unfortunately, we never managed to look into this and, to our knowledge, this is still an open question.

This work showed that proving the formal verification of geometric algorithms is much more involved than one might expect. This led us to explore how other tools could help make the mechanization more practical. One of the outcomes was the integration of QEPCAD [3] and Maple in Isabelle [25, 26, 29] in order to deal with quantifier elimination over the reals and provide tractable manipulations of algebraic expressions. Moreover, these tools provided a way of exploring the iterative construction of loop invariants (and other statements) by easing the discovery of implicit geometric conditions and the removal of contradictory ones. QEPCAD, for instance, allowed the automatic generation of counterexamples that could rule out unprovable geometric conjectures once these were translated into coordinates, while MAPLE enabled the plotting and visualisation of problems as a way of assisting the verification process. By enriching the palette of tools, Laura Meikle went on to tackle much more complicated examples such as the Delaunay triangulation, which involves nested WHILE loops [24].

A natural question, given the difficulties, might be: why should we go to such lengths to reason about geometric algorithms? One answer is that geometric notions underly many of the operations done by autonomous vehicles, e.g., convex hull computations, and membership tests on images are needed to identify regions of interest (such as other cars on the road) [18] and so it is highly likely that providing formal safety guarantees will require formal reasoning about such concepts. Even in less constrained environments, dealing with convex hulls has been shown in later, independent work to be needed when formally proving the correctness of collision avoidance algorithms [46].

5.4 Conclusion

Much of the research described above would not have been possible without the nurturing environment offered by the DREAM Group and its supremo, Alan Bundy. It is difficult to encapsulate in a few pages everything that owes a debt, in some form or another, to the DREAMers and their support. In fact, several research areas, such as process modelling [34] and healthcare [33], that I am currently involved in have not been mentioned. The good news is though: there is much more for us DREAMers to write about in the next book, as we continue to turn dreams into reality.

5.5 Epilogue

Upon arrival in Edinburgh, I found out that I would be moving into Alan Bundy's office as he took on the role of Head of Department and moved to a grander locale. His books would stay behind though as they constituted a valuable DREAM Group resource (and he expected to come back to the office after he finished his term as Head). As a book lover, I was delighted to hear this although it did feel like a big responsibility. However, I would like to believe that I did a good job at the time because Alan asked me a few years ago to look after the books once again. So, it is as the proud Custodian of the Bundy Library, a timeless and priceless DREAM Group asset, that this DREAMer finishes the current account of some of his research.

Acknowledgments This work was supported through grants funded by the Engineering and Physical Sciences Research Council (EPSRC).

Beyond the people mentioned in this chapter, I wish to express my gratitude to all the DREAMers I have had the pleasure of interacting with over the years. Thank you for helping me turn my ϵ-baked dreams into real proofs and tools.

References

1. Yves Bertot and Pierre Castéran. *Interactive Theorem Proving and Program Development - Coq'Art: The Calculus of Inductive Constructions*. Texts in Theoretical Computer Science. An EATCS Series. Springer, 2004.
2. Richard J. Boulton, Konrad Slind, Alan Bundy, and Michael J. C. Gordon. An interface between Clam and HOL. In *Theorem Proving in Higher Order Logics, 11th International Conference*, volume 1479 of *Lecture Notes in Computer Science*, pages 87–104. Springer, 1998.
3. Christopher Brown. QEPCAD-B: a program for computing with semi-algebraic sets using CADs. *ACM SIGSAM Bulletin*, 37:97–108, 01 2003.
4. Alan Bundy, David Basin, Dieter Hutter, and Andrew Ireland. *Rippling: Meta-Level Guidance for Mathematical Reasoning*. Cambridge University Press, 2005.
5. Shang-Ching Chou, Xiao-Shan Gao, and Jing-Zhong Zhang. A collection of 110 geometry theorems and their machine proofs based on full-angles. Technical Report 94-4, CS Dept. WSU, Nov 1994.
6. Shang-Ching Chou, Xiao-Shan Gao, and Jing-Zhong Zhang. Automated generation of readable proofs with geometric invariants, II. Theorem proving with full-angles. *Journal of Automated Reasoning*, 17:349–370, 1996.
7. Christophe Dehlinger, Jean-François Dufourd, and Pascal Schreck. Higher-order intuitionistic formalization and proofs in Hilbert's elementary geometry. In *ADG '00: Revised Papers from the Third International Workshop on Automated Deduction in Geometry*, volume 2061, pages 306–324, London, UK, 2001. Springer-Verlag.
8. Louise A. Dennis and Alan Smaill. Ordinal arithmetic: A case study for rippling in a higher order domain. In *Theorem Proving in Higher Order Logics, 14th International Conference*, volume 2152 of *Lecture Notes in Computer Science*, pages 185–200. Springer, 2001.
9. Lucas Dixon. *A Proof Planning Framework for Isabelle*. PhD thesis, University of Edinburgh, 2006.
10. Lucas Dixon and Jacques D. Fleuriot. IsaPlanner: A prototype proof planner in Isabelle. In *Automated Deduction - CADE-19, 19th International Conference on Automated Deduction*, volume 2741 of *Lecture Notes in Computer Science*, pages 279–283. Springer, 2003.
11. Lucas Dixon and Jacques D. Fleuriot. Higher order rippling in IsaPlanner. In *Theorem Proving in Higher Order Logics, 17th International Conference*, volume 3223 of *Lecture Notes in Computer Science*, pages 83–98. Springer, 2004.
12. Lucas Dixon and Jacques D. Fleuriot. A proof-centric approach to mathematical assistants. *J. Appl. Log.*, 4(4):505–532, 2006.
13. Jacques Fleuriot. *A Combination of Geometry Theorem Proving and Nonstandard Analysis with Application to Newton's Principia*. Springer, 2001.
14. R.L. Graham. An efficient algorithm for determining the convex hull of a finite planar set. *Information Processing Letters*, 1(4):132 – 133, 1972.
15. John Harrison. HOL Light: a Tutorial Introduction. In *Proceedings of the First International Conference on Formal Methods in Computer-Aided Design*, volume 1166, pages 265–269. Springer-Verlag, 1996.
16. Jane Hesketh. *Using middle-out reasoning to guide inductive theorem proving*. PhD thesis, University of Edinburgh, 1992.
17. David Hilbert. *Foundations of Geometry*. Open Court Classics, 10th edition, 1971.
18. Jinkyu Kim and John Canny. Interpretable learning for self-driving cars by visualizing causal attention. In *ICCV: International Conference on Computer Vision*, pages 2961–2969, 10 2017.
19. Ulrich Kortenkamp. Foundations of dynamic geometry. *Journal für Mathematikdidaktik*, 21:161–162, 01 2000.
20. Eleftherios Koutsofios and Stephen North. Drawing graphs with dot. Technical report, AT&T Bell Laboratories, Murray Hill, NJ, 1993.

21. E. Maclean, J. Fleuriot, and A. Smaill. Proof-planning non-standard analysis. In *Proceedings of the 7th International Symposium on Artificial Intelligence and Mathematics*, 2002.
22. Ewen Maclean. *Using proof-planning to investigate the structure of proof in non-standard analysis*. PhD thesis, University of Edinburgh, 2004.
23. David C J Matthews. The Poly and Poly/ML distribution. Technical Report 161, University of Cambridge, Computer Laboratory, February 1989.
24. Laura Meikle. *Intuition in formal proof: a novel framework for combining mathematical tools*. PhD thesis, University of Edinburgh, 2014.
25. Laura Meikle and Jacques Fleuriot. Combining Isabelle and QEPCAD-B in the Prover's Palette. In *Intelligent Computer Mathematics*, Lecture Notes in Artificial Intelligence, pages 315–330, 07 2008.
26. Laura Meikle and Jacques Fleuriot. Prover's Palette: A user-centric approach to verification with Isabelle and QEPCAD-B. In *Computer Aided Verification*, volume 5123 of *Lecture Notes in Computer Science*, pages 309–313, 07 2008.
27. Laura I. Meikle and Jacques D. Fleuriot. Formalizing Hilbert's Grundlagen in Isabelle/Isar. In *Theorem Proving in Higher Order Logics*, volume 2758, pages 319–334. Springer, 2003.
28. Laura I. Meikle and Jacques D. Fleuriot. Mechanical Theorem Proving in Computational Geometry. In *Automated Deduction in Geometry*, pages 1–18, 2004.
29. Laura I. Meikle and Jacques D. Fleuriot. Integrating systems around the user: Combining Isabelle, Maple, and QEPCAD in the prover's palette. *Electron. Notes Theor. Comput. Sci.*, 285:115–119, 2012.
30. Tobias Nipkow. Programming and proving in Isabelle/HOL. https://isabelle.in.tum.de/doc/prog-prove.pdf.
31. Tobias Nipkow. Hoare logics in Isabelle/HOL. In H. Schwichtenberg and R. Steinbrüggen, editors, *Proof and System-Reliability*, pages 341–367. Kluwer, 2002.
32. Joseph O'Rourke. *Computational Geometry in C*. Cambridge University Press, USA, 2nd edition, 1998.
33. Petros Papapanagiotou and Jacques D. Fleuriot. Formal verification of collaboration patterns in healthcare. *Behav. Inf. Technol.*, 33(12):1278–1293, 2014.
34. Petros Papapanagiotou and Jacques D. Fleuriot. WorkflowFM: A logic-based framework for formal process specification and composition. In *Automated Deduction - CADE 26 - 26th International Conference on Automated Deduction*, volume 10395 of *Lecture Notes in Computer Science*, pages 357–370. Springer, 2017.
35. Petros Papapanagiotou, Jacques D. Fleuriot, and Sean Wilson. Diagrammatically-driven formal verification of web-services composition. In *Diagrammatic Representation and Inference - 7th International Conference*, volume 7352 of *Lecture Notes in Computer Science*, pages 241–255. Springer, 2012.
36. Lawrence C. Paulson. *Isabelle: a Generic Theorem Prover*. Number 828 in Lecture Notes in Computer Science. Springer, 1994.
37. Julian Richardson, Alan Smaill, and Ian Green. System description: Proof planning in higher-order logic with $\lambda Clam$. In *Automated Deduction, CADE-15*, volume 1421 of *Lecture Notes in Computer Science*, pages 129–133, 02 1998.
38. Jürgen Richter-Gebert and Ulrich Kortenkamp. *The Interactive Geometry Software Cinderella*. Springer, 1999.
39. Abraham Robinson. *Non-standard Analysis*. Princeton University Press, 1974.
40. Phil Scott. Mechanising Hilbert's *Foundations of Geometry* in Isabelle. Master's thesis, University of Edinburgh, 2008.
41. Phil Scott. *Ordered geometry in Hilbert's Grundlagen der Geometrie*. PhD thesis, University of Edinburgh, 2015.
42. Phil Scott and Jacques D. Fleuriot. Composable discovery engines for interactive theorem proving. In *Interactive Theorem Proving*, volume 6898 of *Lecture Notes in Computer Science*, pages 370–375. Springer, 2011.

43. Phil Scott and Jacques D. Fleuriot. An investigation of Hilbert's implicit reasoning through proof discovery in idle-time. In *Proceedings of the 8th International Conference on Automated Deduction in Geometry*, Lecture Notes in Computer Science, pages 182–200, Berlin, Heidelberg, 2011. Springer.
44. Phil Scott and Jacques D. Fleuriot. A combinator language for theorem discovery. In *Intelligent Computer Mathematics - 11th International Conference*, volume 7362 of *Lecture Notes in Computer Science*, pages 371–385. Springer, 2012.
45. Jamie Stark and Andrew Ireland. Invariant discovery via failed proof attempts. In *Logic Programming Synthesis and Transformation, 8th International Workshop*, volume 1559 of *Lecture Notes in Computer Science*, pages 271–288. Springer, 1998.
46. Holger Täubig, Udo Frese, Christoph Hertzberg, Christoph L uth, Stefan Mohr, Elena Gorbachuk, and Dennis Walter. Guaranteeing functional safety: Design for provability and computer-aided verification. *Autonomous Robots*, 32:303–331, 04 2012.
47. Makarius Wenzel. Asynchronous proof processing with Isabelle/Scala and Isabelle/jEdit. *Electron. Notes Theor. Comput. Sci.*, 285:101–114, 2012.
48. Markus Wenzel. Isar - A generic interpretative approach to readable formal proof documents. In *Theorem Proving in Higher Order Logics, 12th International Conference*, volume 1690 of *Lecture Notes in Computer Science*, pages 167–184. Springer, 1999.
49. Hermann Weyl. David Hilbert and his mathematical work. *Bulletin of the American Mathematical Society*, 50:635, 1944.
50. Freek Wiedijk. The De Bruijn factor. https://www.cs.ru.nl/~freek/factor/factor.pdf, 2000.
51. Freek Wiedijk. Mizar Light for HOL Light. In Richard J. Boulton and Paul B. Jackson, editors, *Theorem Proving in Higher Order Logics, 14th International Conference*, volume 2152 of *Lecture Notes in Computer Science*, pages 378–394. Springer, 2001.
52. Sean Wilson. *Supporting dependently typed functional programming with proof automation and testing*. PhD thesis, University of Edinburgh, 2011.
53. Sean Wilson and Jacques D. Fleuriot. Geometry Explorer: Combining Dynamic Geometry, Automated Geometry Theorem Proving and Diagrammatic Proofs. In *Proceedings of UITP 2005 (User Interfaces for Theorem Provers)*, Apr 2005.
54. Sean Wilson, Jacques D. Fleuriot, and Alan Smaill. Automation for dependently typed functional programming. *Fundam. Inform.*, 102(2):209–228, 2010.

Chapter 6
Reasoned Modelling: Harnessing the Synergies Between Reasoning and Modelling

Gudmund Grov, Andrew Ireland, and Maria Teresa Llano

Abstract Conventional formal modelling requires a designer to have expertise in formal reasoning as well as design. We describe an approach to formal modelling called *reasoned modelling* that aims to allow the designer to focus on their design, with the low-level formal reasoning hidden from view. The approach builds directly upon the ideas of proof plans in that we make explicit use of modelling knowledge and patterns. This enables us to harness the synergies that exist between modelling and reasoning. A number of aspects of reasoned modelling have been investigated. Here we summarise the key contributions that have been previously published. First, when faced with low-level reasoning failures, we illustrate how modelling knowledge can be used to constrain the search for high-level design guidance. Second, we describe how common patterns of refinement can be used to help guide a designer. Third, we outline how common patterns of modelling can be used in suggesting design abstractions. Finally, as is the case with proof plans, reasoned modelling requires a mechanism for instantiating patterns. We describe how automated theory formation was used to instantiate patterns that arose within reasoned modelling.

6.1 Introduction

The use of formal modelling and analysis in the design of complex systems brings significant benefits as well as challenges. Working with a formal design model

G. Grov
Norwegian Defence Research Establishment (FFI), Kjeller, Norway
e-mail: Gudmund.Grov@ffi.no

A. Ireland (✉)
Heriot-Watt University, Edinburgh, UK
e-mail: A.Ireland@hw.ac.uk

M. T. Llano
Monash University, Melbourne, Australia
e-mail: Teresa.Llano@monash.edu

G. Michaelson (eds.), *Mathematical Reasoning: The History and Impact of the DReaM Group*, https://doi.org/10.1007/978-3-030-77879-8_6

allows for strong correctness guarantees to be established early on within the development of a system. In contrast, when using informal modelling notions, such rigour is not possible, and critical design errors may go undetected until the system has been implemented or deployed. On the road to establishing such guarantees, one typically encounters failure, where the process of failure analysis will often give insights into how to progress the design process. Such failures manifest themselves as unproven proof obligations (POs). A broad range of factors may give rise to such failures, e.g., bad design decisions and inconsistencies at the level of the specification. The challenge comes in relating failures at the level of proof obligations to higher-level design decisions and specifications. This is a challenge because it requires the designer to have expertise in both design and formal reasoning.

Our approach to addressing this challenge is to combine both the formal reasoning and modelling aspects in such a way that design and specification are the sole focus of a designer. This involves maintaining a link between the logic and the modelling so that proof failures can be automatically presented in terms of errors at the level of design and specification.

The proof planning approach to theorem proving provides the inspiration for our approach. Having an explicit representation of a common pattern of proof provides guidance in the search for proof instances. But crucially, when failure is encountered in the proof search, the constraints of the pattern can be used in overcoming the failure, e.g., conjecture generalisation and lemmas discovery. Proof planning for automatically discharging the proof obligations arising from formal models was the topic of Yuhui Lin's PhD thesis [27, 28].

A key difference, however, between proof planning and what we call *reasoned modelling* is the need for interaction as well as automation. That is, in addressing a proof failure, we are typically not interested in intermediate lemmas or generalisations—we simply want to know if a conjecture is a theorem. However, proof-failure analysis may also suggest changes to the conjecture. In the case for formal modelling, such changes may affect the design of a system. As a consequence, such changes must be presented to as design alternatives, with the designer in full control of the decision-making.

Figure 6.1 gives a high-level depiction of how user interaction changes with reasoned modelling. On the left, we see that a designer needs expertise to guide both the formal reasoning and the formal modelling aspects of the development, while on the right we see that the goal of reasoned modelling is to reduce this interaction to the modelling aspect only.[1] An anecdote to this change in user interaction can be found for program verifiers. In modern verifiers, such as Dafny [26] and Spark 2014[2], the user will work with the program text only, while the more traditional approach entailed working with a theorem prover in addition to the program text.

[1]Even in the presence of fully automatic provers only, a designer will still need to be able to analyse proof failure in order to determine how the models will need to change.

[2]See https://www.adacore.com/about-spark.

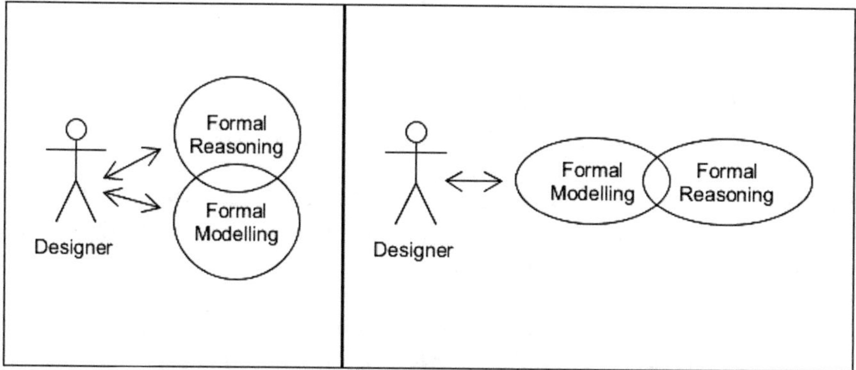

Fig. 6.1 User interaction for posit-and-prove (left) and augmented with reasoned modelling (right)

This chapter provides the first comprehensive presentation of reasoning modelling and is directly based on a corpus of published work and theses [12, 13, 19, 21, 25, 29, 32].

6.2 Refinement-Based Development and Event-B

In this chapter, we focus on a layered style of formal modelling, where a design is developed as a series of abstract models—level-by-level concrete details are progressively introduced via provably correct *refinement* steps.

Event-B [2] is an example of a formal framework following this style of modelling and is mechanised through the Rodin toolset [1]. Here, each step of a development is underpinned by formal reasoning. As a result, there is strong interplay between modelling and reasoning—partly supported by the Rodin toolset. This interplay requires skilled user interaction, i.e., typically a user will analyse failed proofs and translate the analysis by hand into corrective actions at the level of modelling. This is exemplified in [2, 4]. Typical corrective actions include strengthening invariants and guards or modifying actions.

Reasoned modelling aims to provide high-level decision support, by automating the generation, filtering and ranking of modelling suggestions. Event-B models and POs are closely aligned [2], while Rodin [1] is an extensible framework. Event-B and the Rodin toolset thus represent a unique opportunity for us to investigate reasoned modelling.

An Event-B development is structured into *machines* and *contexts*. A context describes the static part of a system, e.g., *constants* and their *axioms*, while a machine describes the dynamic part. Machines are themselves composed of three components: *variables*, *events* and *invariants*. Variables represent the state of the system, events are guarded actions that update the variables and invariants are constraints on the variables. We will use the term *models* for both Event-B contexts and machines. The most basic events are

$$\textbf{EVENT}\ \langle name\rangle \mathrel{\widehat{=}} \textbf{BEGIN}\ \langle action\rangle\ \textbf{END}$$
$$\textbf{EVENT}\ \langle name\rangle \mathrel{\widehat{=}} \textbf{WHEN}\ \langle guard\rangle\ \textbf{THEN}\ \langle action\rangle\ \textbf{END}$$

where the event's action is only executed when the guard holds. INITIALISATION is a special event without guards defining the initial state. In addition, we will see examples of the following event patterns:

$$\textbf{EVENT}\ \langle name_C\rangle \mathrel{\widehat{=}} \textbf{REFINES}\ \langle name_A\rangle\ \ldots$$
$$\textbf{EVENT}\ \langle name\rangle \mathrel{\widehat{=}} \ldots\ \textbf{ANY}\ \langle var_1\rangle\ \ldots \langle var_1\rangle\ \ldots$$

The first event is an example of refinement where the event refines another event (of the abstract model being refined). The second event illustrates how (among other things) arguments to an event can be modelled in Event-B.

6.3 Reasoned Modelling Critics[3]

For any creative activity, understanding our failures often plays a pivotal part in achieving success. This was the motivation for introducing the notion of a *proof critic* [16], a mechanism that supports the analysis and patching of failed proof attempts within the context of proof planning [17]. A proof critic typically exploits partial success in the application of a proof plan. Moreover, it uses partial success to bridge the gap between failed proof attempt(s) and a complete proof. As described by Alan Bundy in his chapter, proof critics have been applied successfully to the problems of inductive lemmas discovery and conjecture generalisation [3, 17, 18], along with the related problem of loop invariant discovery [20]. Proof critics have also been developed for patching faulty conjectures using abduction [31].

Within the context of classical formal verification, a verification task is reduced to a set of purely logical statements. This represents a very powerful divide-and-conquer strategy. However, as a consequence, proof plans and critics cannot take into account the context in which a verification task arose. In terms of proof-failure analysis, this can be problematic. To illustrate, consider an invariant of the form:

$$P \Rightarrow Q$$

Now consider the situation where the verification of the corresponding proof obligation fails. An analysis at the level of atomic propositions can be used to determine how the failure can be overcome. That is, how can we make antecedent

[3]This section is based on [19, 21].

P false or how can we make consequent Q true? To achieve such effects requires changes to the context in which the proof obligation arose—the system and its environment. Armed with such contextual knowledge, it is then possible to rule-out changes that are infeasible, e.g., changes that violate safety constraints, or are impossible, e.g., changes that violate physical constraints. Moreover, certain aspects of a system may be more constrained than others, e.g., safety features. Only a domain expert and/or designer can make such judgements. What we are arguing is that if such knowledge is explicitly represented within the design model, then it can be used to automatically filter modelling changes when a verification attempt fails. This led us to consider how the ideas underpinning the notion of a proof critic evolved so as to combine knowledge of the domain being modelled as well as knowledge of proof—to give the notion of a *reasoned modelling critic*.

To illustrate these ideas, we will consider a simple cruise control system where the variable *brake* is used to represent the state of the brakes and *cc* is used to represent the state of the cruise controller. Initially, both variables are *off* (where the || operator denotes the actions within the initialisation event being executed in parallel):

$$\text{INITIALISATION} \mathrel{\widehat{=}} \textbf{BEGIN}\ \mathit{brake} := \mathit{off} \mathbin{||} \mathit{cc} := \mathit{off}\ \textbf{END}$$

We will focus on two events associated with the system. The first occurs when the cruise controller is enabled:

$$\textbf{EVENT}\ \text{enable\_cc} \mathrel{\widehat{=}} \textbf{BEGIN}\ \mathit{cc} := \mathit{on}\ \textbf{END}$$

The second occurs when a driver presses the brakes, as defined by the event:

$$\textbf{EVENT}\ \text{pressbrake} \mathrel{\widehat{=}} \textbf{BEGIN}\ \mathit{brake} := \mathit{on}\ \textbf{END}$$

A key safety constraint associated with the system is that the brakes (*brake*) cannot be on, while the cruise controller (*cc*) is enabled. Logically, this can be formalised as an invariant of the form:

$$\mathit{cc} = \mathit{on} \Rightarrow \mathit{brake} = \mathit{off}$$

Now turning to the verification proofs. The enable_cc event gives rise to a proof obligation of the form (where $\{x \mapsto y\}T$ denotes a substitution of x to y in T):

$$\mathit{cc} = \mathit{on} \Rightarrow \mathit{brake} = \mathit{off} \vdash \{\mathit{cc} \mapsto \mathit{on}\}(\mathit{cc} = \mathit{on} \Rightarrow \mathit{brake} = \mathit{off}) \tag{6.1}$$

which can be reduced to the unprovable goal:

$$\mathit{cc} = \mathit{on} \Rightarrow \mathit{brake} = \mathit{off} \vdash \mathit{brake} = \mathit{off}.$$

Similarly, the pressbrake event creates the following proof obligation:

$$cc = on \Rightarrow brake = off \vdash \{brake \mapsto on\}(cc = on \Rightarrow brake = off) \qquad (6.2)$$

which, when simplified, becomes false:

$$cc = on, brake = off \vdash false$$

In terms of overcoming these proof failures, we focus here upon changes to the associated events. Specifically, how the addition of actions and/or guards will lead to successful proofs.

The failure to prove (6.1) is associated with the event enable_cc. Proof-failure analysis tells us that we can overcome the failure by making the consequent $brake = off$ true. At the modelling level, this can be achieved by the addition of an action:

$$\textbf{EVENT}\ \text{enable\_cc} \mathrel{\widehat{=}} \textbf{BEGIN}\ cc := on\ ||brake := off\ \textbf{END}$$

Alternatively, we can achieve the same effect by the addition of a guard:

$$\textbf{EVENT}\ \text{enable\_cc} \mathrel{\widehat{=}} \textbf{WHEN}\ brake = off\ \textbf{BEGIN}\ cc := on\ \textbf{END}$$

Note that the first alternative change results in the brakes being released, i.e., *brake* is set to *off*, as a side effect of enabling of the cruise controller (enable_cc). While this overcomes the proof failure, it is obviously an unsafe event. In contrast, the second alternative change ensures that the cruise controller (enable_cc) can only be enabled when the brake has not been applied, i.e., $brake = off$. The desirable solution requires the introduction of the guard that ensures that the cruise controller cannot override the brakes.

Now consider the failure to prove (6.2), which is associated with the event pressbrake. Proof-failure analysis tells us that we can overcome the failure by making the antecedent $cc = on$ false. At the modelling level, this can be achieved by the addition of an action:

$$\textbf{EVENT}\ \text{pressbrake} \mathrel{\widehat{=}} \textbf{BEGIN}\ brake := on\ ||cc := off\ \textbf{END}$$

or alternatively, we can achieve the same effect by introducing a guard:

$$\textbf{EVENT}\ \text{pressbrake} \mathrel{\widehat{=}} \textbf{WHEN}\ cc = off\ \textbf{BEGIN}\ brake := on\ \textbf{END}$$

Here, the first alternative change results in the cruise controller being disabled, i.e., *cc* is set to *off*, as a side effect of the brake being pressed (pressbrake). In contrast, the second alternative change ensures that the brake can only be pressed when the cruise controller has been disabled, i.e., $cc = off$. The desirable solution requires the introduction of the action that ensures that the cruise controller cannot override the brakes.

As highlighted earlier, knowledge of the domain being modelled is required in order to identify desirable solutions. For the given example, being able to brake is clearly more important than driving with the cruise control enabled. This suggests that certain variables within a model have higher *priority* than others. Moreover, this notion of priority can be used to heuristically rank, or even prune, certain undesirable changes:

H1: If the priority of the candidate variable is lower than the priorities of all the variables updated by the event, then it is strongly suggestive that the change should be achieved via a new action.
H2: If the priority of the candidate variable is higher than the priorities of all the variables updated by the event, then it is strongly suggestive that the change should be achieved via a new guard.

As emphasised earlier, domain knowledge, such as the priority of variables that occur within model, must be supplied by the designer or domain expert.

While clearly a very simple example, it illustrates how combining domain knowledge with proof-failure analysis can deliver more relevant guidance to a designer. More details on the modelling critics ideas can be found in [19]. The paper describes how the critic frame can be extended to combine both models and proof obligations. It also discusses the potential for generating more general modelling changes by combining the analysis of multiple proof failures.

6.4 Refinement Plans[4]

A proof critic handles a special pattern of failure (or partial matches) of a proof plan. Analogically, a modelling critic should handle a partial match of a "modelling plan". In Event-B, a common approach to developing formal models is through step-wise refinement, and for this type of modelling, we have captured a set of "modelling plans" that we have called *refinement plans*.

By analysing a range of Event-B case studies from the literature, eight basic refinement plans have been identified and grouped into a hierarchy as shown in Fig. 6.2. Each refinement plan contains patterns of refinement identified by syntactic features of abstract and concrete models, as well as patterns of the underlying POs to justify the refinement. Modelling critics can then be associated to partial matches of these patterns.

The classification from Fig. 6.2 provides us with a better understanding of what a user is trying to achieve in a refinement step as well as facilitates the matching process. We first briefly summarise each refinement plan before illustrating use of one of them:

[4]This section is based on [12].

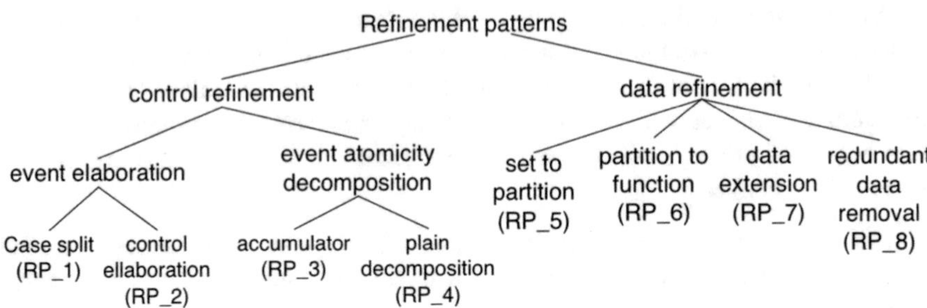

Fig. 6.2 Hierarchy of refinement patterns

Case split: refers to refinement steps in which an abstract event is refined in the concrete model by two or more events.

Control elaboration: relates to models that constrain the application of existing events based on extensions of the state and independently from the operation of new events at the concrete level.

Accumulator: deals with models in which actions of an abstract atomic event are performed in the concrete model via iteration.

Plain decomposition: makes reference to models in which an abstract event is refined by a sequence of new and refined events. New events are used to pre-process data used in the abstract event.

Set to partition: refers to models in which an abstract variable is refined by partitioning it through a set of new variables in the concrete model.

Partition to function: involves refinement steps in which an abstract partition of variables is refined into a function in the concrete model.

Data extension: refers to models in which an abstract variable is refined into a concrete variable that extends the abstract data type in order to control membership of data in the variable.

Redundant data removal: involves the elimination of data from the abstract level that is not being used to control the operation of any event.

We will only focus on the *accumulator refinement plan* (RP_3) and two of its associated critics. This refinement plan has its roots in earlier work on accumulators, both in recursive and iteration program verification, and intuitively deals with models in which actions of an abstract atomic event are performed in the concrete model via iteration. This iteration is achieved through the use of new events that iteratively accumulate the value from the abstract action, and the plan has been taken from work by Butler and Yadav [4], which was further developed in [5, 9, 10].

The key difference with our work compared with Butler et al is that as well as the modelling patterns, we are also interested in the deductive patterns and in providing guidance when a pattern breaks in a development.

We illustrate the plan with a simple model taken from [8], where an event *incr* adds a value y to a variable x:

EVENT incr $\widehat{=}$ **BEGIN** x := x + y **END**

The concrete model, which refines this abstract model, has a set of *accumulator variables* and three events:

```
EVENT start ≙
 WHEN
  flag = TRUE
 THEN
  n := 0
  x_tmp := x
  flag := FALSE
END
```

```
EVENT step ≙
 WHEN
  n < y
  flag = FALSE
 THEN
  x_tmp := x_tmp + 1
  n := n + 1
END
```

```
EVENT end_ok ≙
REFINES incr
 WHEN
  flag = FALSE
 THEN
  x := x_tmp
  flag := TRUE
END
```

Event *start* is the initialisation event, event *step* is an "accumulator event" and *end_ok* is a "refining event". A new counter *n* is initialised to 0 by event *start*, and this is used by event *step* to iteratively assign the value of *y* via an accumulator variable *x_tmp*. At the end of the iteration, event *end_ok* assigns the value of *x_tmp* to (the abstract) variable *x*.

The pattern requires an invariant that explains the refinement, meaning that the content of the accumulator variable(s) is contained within the value assigned in the abstract model. The main properties are that initialisation, accumulator(s) and refined events must preserve the invariant and that the refined event must preserve the behaviour of the abstract event.

For the example, the invariant is missing and the corresponding proofs fail; thus, the example illustrates a flawed, or partly matching, instance of the pattern. To overcome such a failure, a set of critics are defined, and we will illustrate two such critics for the accumulator plan:

postGuard_speculation critic: considers the case when the guard of the refined event that ensures the accumulation process is complete is either flawed or missing.

invariant_speculation critic: handles the case when the accumulator invariant is wrong or missing.

The critics are triggered by combining partial matching of the modelling patterns with failure analysis. This is used to automatically generate modelling guidance. Both the above critics are applicable in this example.

The postGuard_speculation critic will add a guard with the shape of an equality to the refined event. In our case, this event is *end_ok* and the critic will use the variables *x_tmp,n, x,y* to instantiate the guard, which will result in an additional guard $y = n$ being added to the event.

This moves the model to be closer to correct, but since the invariant is also missing, the failure persists even after adding this guard—thus triggering the invariant_speculation critic. It will speculate an invariant of the shape $H_1 \Rightarrow F_{\preceq}(\overline{W}, V_2)$, where $F_{\preceq}$ denotes an inequality, which will be added to the concrete model. In our case, this is

$$(flag = FALSE) \Rightarrow F_{\preceq}(x\_tmp, n, x, y)$$

which is instantiated to

$$flag = FALSE \Rightarrow x\_tmp = x + n$$

This completes the refinement.

It is important to note that the critics provide guidance in the form of partially instantiated schemas. The process of completing the instantiation has been tackled by a range of techniques that have been developed within the DReaM Group over a number of years.

First, in his chapter, Alan Bundy illustrated how meta-variables can be used to specify intermediate lemmas and how the constraints of rippling can be used to instantiate the specification. This approach, often referred to as *middle-out reasoning*, has also been applied to the problems of conjecture generalisation and loop invariant discovery. Higher-order unification played a central part of the technique, with the rippling annotations being used to divide the unification task into smaller tasks. However, in general, the constraints of rippling were not sufficient and required the piece-wise unification needed to be interleaved with a process of projection and conjecture disproving [17].

Second, a radical and highly successful alternative is the bottom-up generate-and-test approach known as *term synthesis* [24, 30].

Here, we investigated yet another radically different approach to discovering unknowns—automated theory formation (ATF). Specifically, we automated the instantiation of invariant patterns within refinement plans by applying ATF to simulation traces of our Event-B design models. Our application of ATP was embedded in a tool called HREMO, which is described in the next section.

6.5 Invariants Generation and HREMO[5]

Refinement plans provide instantiations of well-known modelling patterns; however, the partial and static nature of these instantiations limited their application to known types of failures. One of the tasks faced by developers when a refinement step fails is to supply invariants that relate to their design decisions. Following one of the DReaM Group's motifs on the productive use of failure, we developed a heuristic approach that supports the activity of formal modelling, and complemented

[5] This section is based on [32].

Variables	**Invariants**	**Events**
full	full ∈ BOOL	**INITIALISATION** $\widehat{=}$ **BEGIN** full := false **END**
		EVENT addA $\widehat{=}$ **WHEN** full = false **BEGIN** full :∈ BOOL **END**

(a) Abstract Level

Variables	**Invariants**	**Events**
x m	x ∈ ℕ	**INITIALISATION** $\widehat{=}$ **BEGIN** x := 0 ‖ m := 3 **END**
	m ∈ ℕ	**EVENT** addC $\widehat{=}$ **REFINES** addA **WHEN** x < m
		WITH full=false ⇔ x<m **BEGIN** x := x+1 **END**

(b) Concrete Level

Fig. 6.3 Flawed Event-B model

refinement patterns, by automatically discovering invariants. Briefly, the approach requires:

1. A formal modelling component that supports proof
2. A simulation component that generates system traces
3. Automated theory formation (ATF), a technique that identifies patterns from examples of a background domain, which generates conjectures from the analysis of the traces

To illustrate this, observe the model in Fig. 6.3. At the abstract level, the Boolean variable *full* is modified in event *addA* through a non-deterministic action when *full* is false. Note that the non-deterministic action, i.e., *full* :∈ *BOOL*, specifies that variable *full* is non-deterministically assigned a Boolean value. At the concrete level, the state of the system is refined by replacing the abstract variable *full* by concrete variables x and m. Moreover, the abstract event *addA* is refined by the concrete event *addC*, which gradually increments variable x by one unit when x is less than m.

As it stands, the model generates a PO to capture correct strengthening of the guard. This fails to prove as shown in (6.3). This PO specifies that the concrete guard $x < m$ must imply the abstract guard $full = false$.

$$x < m \vdash full = false \tag{6.3}$$

The failure is generated because the relationship between the abstract and concrete states, captured by a special type of invariant called a *gluing invariant*, has not been defined.

We use ATF, and specifically the HR system [6, 7], to search for the missing invariant. HR forms theories about a domain through an iterative application of general-purpose production rules (PRs) for concept invention. PRs are either unary or binary; this means that a PR must be applied to either one or two input data tables (depending on their operation). Data tables represent concepts by means of a set of examples, and applying a unary or binary PR results in a new data table, i.e., a new concept. HR then examines whether the new concept is: (i) equivalent to an existing concept; (ii) subsumed by or subsumes an existing concept or (iii) empty. These relationships take the form of equivalence, implication, or non-existence conjectures, respectively.

Fig. 6.4 Simulation trace generated by the ProB simulator

		Animation steps			
	Variables	**S1**	**S2**	**S3**	**S4**
Abstract	full	false	false	false	true
Concrete	x	0	1	2	3
	m	3	3	3	3

Particular to the invariant discovery process, the first step is to provide information about the domain for which the theory will be formed, i.e., the Event-B model. Specifically, HR requires: (i) user-given concepts that enumerate the objects of interest (we call these T1 concepts) and (ii) user-given concepts that define features of the objects of interest (we call these T2 concepts). Within Event-B, T1 concepts are given by the context (static part), while T2 concepts are formed by variables and constants. For the model (M) in Fig. 6.3, these correspond to

$$\begin{aligned} conceptsT1(M) &= \{boolean, integer\} \\ conceptsT2(M) &= \{full, x, m\} \end{aligned}$$

Additionally, for each concept, HR requires a set of examples in order to apply its PRs. Within the context of Event-B, simulation provides a source of such examples. Through simulation, it is possible to analyse the operation of an Event-B model by observing how its state changes when different scenarios are explored. The simulation trace shown in Fig. 6.4 is produced by the ProB simulator [15] for the flawed Event-B model. The trace shows the value of the abstract and concrete variables at each step of the simulation.

The information from the simulation traces is then transformed into data tables formed of the following tuples: state(A) = $\{\langle S1\rangle, \langle S2\rangle, \langle S3\rangle, \langle S4\rangle\}$, boolean(B) = $\{\langle true\rangle, \langle false\rangle\}$, integer(C) = $\{\langle 0\rangle, \langle 1\rangle, \langle 2\rangle, \langle 3\rangle\}$, full(A,B) = $\{\langle S1, false\rangle$, $\langle S2, false\rangle$, $\langle S3, false\rangle$, $\langle S4, true\rangle\}$, x(A,C) = $\{\langle S1, 0\rangle$, $\langle S2, 1\rangle$, $\langle S3, 2\rangle$, $\langle S4, 3\rangle\}$, m(A,C) = $\{\langle S1, 3\rangle$, $\langle S2, 3\rangle$, $\langle S3, 3\rangle$, $\langle S4, 3\rangle\}$—where A specifies an argument of type $state$, B an argument of type $boolean$ and C an argument of type $integer$. The concept *state* represents every step of the simulation trace, the concepts *boolean* and *integer* are the data types and the concepts *full, x* and *m* are the abstract and concrete variables of the model. We call this set of input concepts *core concepts*.

With this background information, HR begins the theory formation process by applying all possible combinations of concepts and production rules in order to generate new concepts, which we call *non-core concepts*, and identify relationships between them, i.e., the conjectures.

In order to focus the output produced by HR, we apply two types of heuristics: Configuration Heuristics (CH) and Filtering Heuristics (FH). We use the CH heuristics to influence how HR organises the concepts to be explored and to constrain the applicable PRs during theory formation, while the FH heuristics focus in finding conjectures that would address the failures expressed by the POs. The heuristics are as follows (we show the application of the heuristics to the model in Fig. 6.3, and failed PO (6.3)):

Configuration heuristics

CH1: *Prioritise core and non-core concepts that occur within the failed POs*: focusing HR's interest on concepts related to the failures that arose within the model. For example, the application of heuristic CH1 results in the following sets of prioritised core and non-core concepts:

$$\begin{aligned} \textit{prioritised core concepts} &= \{\textit{full}, x, m\} \\ \textit{prioritised non-core concepts} &= \{\textit{full}=\textit{false}, x<m\} \end{aligned}$$

CH2: *Select the subset of PRs that are most relevant to the given failed POs*: selecting in this way PRs that focus HR's theory formation on potential syntactic similarities between the failed POs and the missing invariants. Applying CH2 to the example produces a set of 5 PRs: the *compose*, *disjunct* and *negate* PRs, which are enabled by default for the task of invariant discovery as they perform basic logic operations (conjunction, disjunction and complement, respectively) commonly seen in Event-B invariants; the *split* PR, which allows HR to focus on specific examples of the data table, is selected because of the presence of the value *false* in the goal of the PO (i.e., we will focus on examples of the variable *full* that are set to *false*); and the *numrelation* PR, which performs arithmetic comparisons ($<$, $>$, $\leq$, $\geq$), is selected because of the occurrence of operator $<$ in the hypothesis.

Filtering heuristics

FH1: *Select conjectures that focus on prioritised core and non-core concepts*: In other words, we are interested only in equivalence and implication conjectures where either the left- or right-hand side represents a prioritised concept, as well as in all non-existence conjectures where a prioritised concept occurs. Applying heuristic FH1 to the example results in the selection of 14 conjectures: 2 equivalences, 4 implications and 8 non-existence conjectures.

FH2: *Select conjectures where the sets of variables occurring on the left- and right-hand sides are disjoint*: identifying only those that do not contain multiple occurrences of a variable—typically the set of variables involved in an invariant does not contain duplicates. The application of the heuristic does not reduce the number of conjectures. This may occur because of the simplicity of the model.

FH3: *Select only the most general conjectures*: eliminating redundancies among the selected conjectures by removing those that are logically implied by more general conjectures. Applying this heuristic to the example reduces the set of selected conjectures to 7: 1 equivalence, 3 implications and 3 non-existence conjectures.

FH4: *Select conjectures that discharge the failed POs*: so that only conjectures that help overcome the failures are selected. Applying this heuristic to the example results in the selection of only two conjectures: 1 equivalence and 1 implication.

FH5: *Select conjectures that minimise the number of additional proof failures that are introduced*: since overcoming a proof failure potentially leads to new proof failures. Regarding the example, the two conjectures selected by heuristic FH4 discharge the failed PO and do not produce any extra failure; thus, both of them are presented to the user as candidate invariants:

$$full = FALSE \Leftrightarrow x < m \tag{6.4}$$

$$x \neq m \Rightarrow full = FALSE \tag{6.5}$$

Through manual inspection, conjecture (6.4) is identified as the missing invariant of the flawed model.

As has been observed throughout this section, our heuristic approach exploits the strong interplay between modelling and reasoning in Event-B by using the feedback provided by failed POs to make decisions about how to configure HR. Furthermore, using proof-failure analysis to prune the wealth of conjectures HR discovers, these heuristics have proven highly effective at identifying missing invariants.

6.6 Design-Space Exploration[6]

A drastic alternative to our previously explained ideas of applying proof plans and critics to the modelling level is to apply ATF to this modelling level—which we call *design-space exploration* (DSE).

We will illustrate this approach through a worked example of a simplified protocol for transferring money between bank accounts with the following requirements:

R1: The sum of money across all accounts should remain constant.
R2: Transactions can only be completed if the source account has enough funds.
R3: If an amount m is debited from a source account, the target account should be credited by m.
R4: Progress should always be possible (no deadlocks).

A designer might choose to represent the protocol as follows in Event-B:[7]

EVENT $start \mathrel{\widehat{=}}$ **ANY** $a1\ a2\ m$
  WHEN $a1 \notin active$
  THEN $pend := pend \cup \{((a1, a2), m)\} \parallel active := active \cup \{a1\}$
EVENT $debit \mathrel{\widehat{=}}$ **ANY** $a1\ a2\ m$
  WHEN $((a1, a2), m) \in pend \wedge bal(a1) \geq m$
  THEN $bal(a1) := bal(a1) - m \parallel pend := pend \setminus \{((a1, a2), m)\} \parallel$
    $trans := trans \cup \{((a1, a2), m)\}$
EVENT $credit \mathrel{\widehat{=}}$ **ANY** $a1\ a2\ m$
  WHEN $((a1, a2), m) \in trans$

[6] This section is based on [13, 25].

[7] To ease presentation, we make an implicit assumption that for a given time there is only a single transaction between two accounts. Consequently, *trans* can be treated as a function from a pair of accounts to a value, written $trans(a1, a2) = m$ for $((a1, a2), m) \in trans$.

THEN $bal(a2) := bal(a2) + m \parallel trans := trans \setminus \{((a1, a2), m)\} \parallel$
$active := active \setminus \{a1\}$

The chosen representation involves three steps, each of which is represented through an event that is parametrised by the names of the source ($a1$) and target ($a2$) accounts, along with the value of money (m) associated with the transfer. Step one (event $start$) initiates a transfer by adding the transaction to a *pending* set ($pend$) and uses a set ($active$) to ensure that an account can only be the source of one transfer at a time. The second step (event $debit$) removes the funds from the source account if sufficient funds exist—*bal* denotes a function that maps an account to its balance. If successful, the transaction is removed from the *pending* set and is added to the *transfer* set. The final step (event $credit$) completes the transaction by adding the funds to the target account, as well as updating the $trans$ and $active$ sets accordingly. Finally, requirement R1 is formalised as an invariant,

$$\text{I1: } \Sigma_{a \in dom(bal)} bal(a) = C$$

where C is a constant that represents the sum of money across all accounts.

Key to DSE is *abstraction*—the ability to create a design at the right level of detail and to "glue" it to any abstract model through a set of gluing invariants. Trial-and-error is very much part of the expert methodology, where low-level proof failures are examined, and design alternatives in terms of abstractions are experimented manually (see [4]). Within DSE, our goal is to automate much of the low-level grind associated with the trial-and-error nature of formal modelling and provide a designer with *high-level* modelling advice in real time. In particular, the aim is to generate alternative models at a higher level of *abstraction* than the original model to deal with a flaw. The intuition is that the flaw is a result of being too concrete. Moreover, within a correct abstraction, the designer has the additional burden of correctly defining the system behaviour and supplying numerous auxiliary invariants that are required for the formal verification process. To support this, *adaptations* of the initial model are suggested at the same level of abstraction. This could be for instance in terms of additional invariants, or even changes to the behaviour of the system.

Unconstrained generation of new models will result in an enormous search space that will be infeasible in practice. To overcome this, ATP and HREMO are first utilised to highlight problematic areas in order to pinpoint where the problem lies and what the problem might be. This will be used to explore alternative abstractions and adaptations of the model. These are generated through a set of low-level but generic "atomic operators" that make small changes to a model, e.g., "delete variable" and "merge events". These atomic operators can then be combined in order to generate new models and constrained to reduce the number of possible models generated. Common modelling patterns play a central role in finding the right combination of operators. These will be at a very high level to enable flexibility in terms of their application and therefore enable us to provide assistance in situations where there are no applicable design patterns. Abstraction patterns are essentially inverse application of *refinement patterns*, and we will next illustrate two such patterns:

1. "Undoing" bad behaviour by introducing a special error (or exception) case.
2. Abstracting away the problem when it can be pinpointed between certain events. This amounts to "atomising" sequential events into a single event.

Consider again the user provided model of a money transfer protocol given above. As it stands, the model is flawed since R4 is violated when all accounts have started a transaction but none of the source accounts have sufficient funds. Moreover, event *debit* violates invariant I1 since the amount removed from the source account is not accounted for in the invariant, which breaks requirement R2. Our aim in such situations will be to offer the designer modelling alternatives that address the flaws. Figure 6.5 summarises the alternatives generated through our approach, and below we outline how this was achieved.[8]

(*Abstraction A1*) Applying ATF indicates "bad behaviour" associated with event *debit* and variable *active*. We can apply the "abstract away" pattern to this violation. One implementation of this pattern is to remove the variable that two (sequential) events use to communicate an intermediate result, and then combine this sequence into an atomic event. One (out of two) such combined event involving both *debit* and *active* is

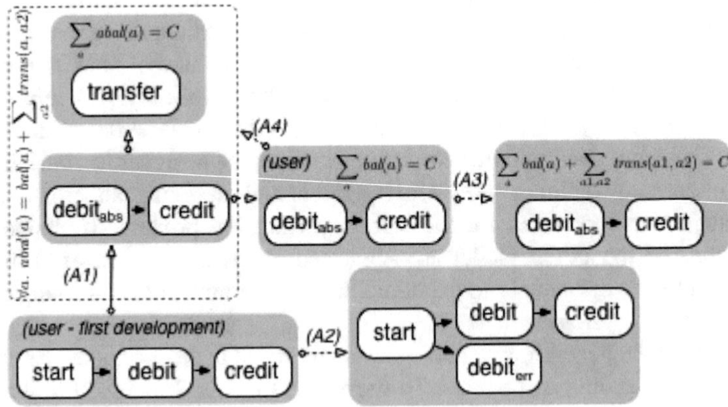

Fig. 6.5 A diagrammatic summary of a small design-space exploration: Starting from the initial development, abstraction (A1) and adaptation (A2) are suggested to deal with violation of requirement R4. Given that I1 is a near-invariant, a new invariant is suggested in (A3) or an abstraction (A4) with the required gluing invariant

[8]More details of the examples are available on arXiv [11].

$$\begin{array}{ll}\textbf{EVENT}\ debit_{abs} \mathrel{\widehat{=}} & \textbf{ANY}\ a1\ a2\ m \\ & \textbf{WHEN}\ a1 \notin active \wedge bal(a1) \geq m \\ & \textbf{THEN}\ active := active \cup \{a1\} \\ & \qquad \parallel bal(a1) := bal(a1) - m \\ & \qquad \parallel trans := trans \cup \{((a1, a2), m)\}\end{array}$$

(*Adaptation A2*) An alternative analysis is to apply the error-case pattern. Intuitively, this means introducing a new "error-handling" event that will "undo" some previous state changes when the desired path is not applicable. This can be implemented so that it reverses a previous action in cases when an event of the desired path stays disabled. This requires transformations to negate an event's guard, reverse an action of an event and combine the guards of one event with the actions of another. One (out of seven) such error-handling events that can be generated, given the constraints w.r.t. *debit* and *active*, is

$$\begin{array}{ll}\textbf{EVENT}\ debit_{err} \mathrel{\widehat{=}} & \textbf{ANY}\ a1\ a2\ m \\ & \textbf{WHEN}\ ((a1, a2), m) \in pend \wedge bal(a1) < m \\ & \textbf{THEN}\ pend := pend \setminus \{((a1, a2), m)\} \\ & \qquad \parallel active := active \setminus \{a1\}\end{array}$$

Event $debit_{err}$ handles the case when the source account does not have enough funds.

(*Adaptation A3*) Let us assume the user selects abstraction *A1*. Through analysis of this alternative, we can see that we are in a "bad state" when *trans* and *active* are not empty, i.e., when there are transactions currently in progress. As a result, ATF will be used to search for conjectures that involve the concepts *trans* and *active* as well as the invariant itself, i.e.,

$$C = \Sigma_{a \in dom(bal)}\ bal(a)$$

ATF is then able to generate an *adaptation* of the invariant I1 that addresses the violation by $debit_{abs}$. The Event-B representation of the invariant, which replaces I1, is

$$\text{I2: } \Sigma_{a \in dom(bal)} bal(a) + \Sigma_{(a1,a2) \in dom(trans)} trans(a1, a2) = C$$

(*Abstraction A4*) Although correct, invariant I2 is not a natural representation of R1, as compared with near-invariant I1. The designer may wish to explore an alternative abstraction in which I1 is an invariant. Our final alternative *A4* represents such an abstraction. Based on the analysis for alternative A1, we can re-apply our "abstract away" pattern, albeit with a slightly modified configuration that deletes two variables. Constrained by the analysis, two such alternatives can be generated with one of them being the desired *transfer* event:[9]

$$\begin{aligned} \textbf{EVENT}\ \mathit{transfer} \mathrel{\widehat{=}} \ & \textbf{ANY}\ a1\ a2\ m \\ & \textbf{WHEN}\ abal(a1) \geq m \wedge a1 \neq a2 \\ & \textbf{THEN}\ abal(a1) := abal(a1) - m\ || \\ & \quad\ ||\ abal(a2) := abal(a2) + m \end{aligned}$$

Finally, in order to prove the consistency between the abstract and concrete models, a gluing invariant is required, where ATF and HREMO are used to form a theory of the refinement step and search for the invariant. HREMO is able to figure out the relation between the abstract variable *abal* and the concrete representation, i.e., variables *bal* and *trans*. A key challenge here will be tailoring HREMO for the formal methods context so that invariants such as the gluing invariant required in this refinement step can be formed.

6.7 Future Work and Conclusion

In this chapter, we have given an overview of our reasoning modelling paradigm—both the overall ideas and specific approaches we have explored. It is also important to note that even if we have focused here on illustration of the concepts through examples, all the ideas presented have been implemented in proof-of-concept prototypes.

While our work on reasoned modelling has been grounded in the Event-B formalism, we believe there is a more generic story to reasoned modelling. We therefore end our discussion by sketching how we are currently developing the core ideas in a range of new directions.

6.7.1 Requirements, Domain Properties and Specifications

As emphasised in Sect. 6.3, failure analysis that focuses purely on logic will typically generate many modelling suggestions that are either infeasible or impossible given the knowledge about the domain of application. While we have focused here on the modelling of discrete systems, and specifically Event-B, the ideas are more widely applicable. For example, while proof failure may be overcome by revising design decisions, it may also require revisions to the assumptions that have been made about the application domain. Alternatively, proof-failure analysis may highlight infeasible system-wide requirements. The *problem frames* approach [23]

[9]To simplify presentation we abuse the Event-B notation slightly here.

to requirements engineering was developed with the aim of making specifications, domain properties and requirements explicit. There are links between proof plans and problem frames. While the former captures common patterns of reasoning, the latter captures common patterns of problem within the context of software systems. We are currently investigating how our modelling critics ideas can be used to support the problem frames methodology.

6.7.2 *Hazard Analysis—What-if Style Scenarios*

Current approaches to safety analysis are limited to the information provided by the requirements and to the capacities of the designers to question the possible failures that can arise. However, complex systems involve different participants whose behaviour is not consistent with the modelled processes, components that are distributed all across the environment and events that are out of the participants control. All these can produce hazardous situations difficult to imagine to the designers of the system. Take for instance the case of the Hull Paragon accident occurred on February 14, 1927[10] where a railway operator changed a signal before time, while another operator pulled the wrong lever, causing two trains travelling in opposite directions to be on the same track and crash. The simultaneous nature of the events, coupled with other human-related factors, had not been previously identified by the designers of the signalling system.

In [22], we studied how dysfunctional behaviour could be identified by relying on meta-knowledge to constrain the search for properties that lead to hazardous situations. This meta-knowledge can be provided by the designer, as proposed in [22], but could also be obtained through other sources of knowledge by exploring how unnatural chains of events can occur within a system. Our work on computational creativity has analysed the use of NLP techniques to semantically manipulate information from knowledge bases in order to generate fictional scenarios. Here, we are interested in applying this type of creative mechanisms for studying how to generate "plausible scenarios" for a given system model. That is, we want to study how we can use information from a system's records (e.g., log files, models' simulations, etc.) in order to generate What-if like scenarios of situations that may produce accidents. For instance, we expect to find scenarios such as: "What if a plane landed on a wet runaway?", "What if a person opened the door of an active washing machine?", "What if a railway operator fell asleep during their shift?", etc. These scenarios could be fed back to the designers and when possible to the model as meta-knowledge, opening up the knowledge about the system and its environment in order to discover possible hazardous scenarios.

[10]http://www.railwaysarchive.co.uk/docsummary.php?docID=308.

6.7.3 Enterprise Security Architecture

Enterprise security architecture (ESA) is a term often used for the security aspects of an enterprise architecture (EA)—a holistic view of an enterprise including how different aspects of the enterprise are related. Such aspects could include everything from overall capabilities, business processes and computer systems. Security is an all-encompassing concern for an enterprise and therefore found in most, if not all, of the aspects. As a result, analysis of security concerns often requires the analyst to simultaneously address several aspects of the enterprise, which is infeasible as a purely manual task. We have therefore started to address automated reasoning in this context [14].

While such formal reasoning is an important step forward, it will still only get us to the left-hand side of Fig. 6.1, where security architects need automated reasoning skills in addition to the required expertise in enterprise modelling and security.

The idea of "hiding" the reasoning and communication at the modelling level only (right-hand side of Fig. 6.1) is just as important here. Moreover, EA exhibits a hierarchical structure between certain aspects, where the higher level provides a form of abstraction of the lower levels. A desirable feature is consistency between these layers, providing a clear link to refinement and refinement plans. Generation of abstractions from the low-level (and typical technical layers) to the high-level (typical business) layers—as in design-space exploration—is another desirable feature for an EA/ESA that would be worth exploring.

6.7.4 Conclusion

As well as a powerful automated reasoning technique, the proof planning paradigm provides a computational framework to investigate and explain proof strategies—*providing a kind of how-to-guide for proving theorems!* As noted above, this led to our ability to automatically analyse and patch a broad range of proof failures through proof critics. Moreover, the explanatory nature inspired us to investigate how one can combine proof and design through reasoned modelling. And as indicated above, this represents a rich vein of future research opportunities.

Acknowledgements and Final Reflections

We are indebted to many past and present members of the group for their support and guidance. We are particularly indebted to Colin O'Halloran (D-RisQ) and Ben Gorry (BAE Systems), both of whom have had a long association with the group and have supported our work in particular. In addition, we would like to thank the reviewers for their detailed and encouraging feedback on a previous version of the paper, and to Greg [Michaelson] for his work on putting together this book.

Finally, turning to the DReaM Group and Alan Bundy's leadership. The group has provided a unique environment for nurturing young researchers for over 40 years. Alan's generosity, creativity and passion for getting to the heart of research problems continues to inspire. He has mentored countless researchers that have gone on to develop highly successful academic careers—for which we owe a huge debt of gratitude. We end the paper with some individual reflections on our relationship to the DReaM research environment.

Andrew I joined the then Department of Artificial Intelligence in October 1988 as a "University Demonstrator". While supporting students on the Knowledge-Based Systems MSc programme, I spent my evenings writing-up my PhD thesis (entitled "*Mechanization of Program Construction in Martin-Löf's Theory of Types*"), which I had undertaken at the University of Stirling. I joined the DReaM Group on January 1st 1990 as a Post-doctoral Research Associate. Before joining the group, I was unsure whether I could build an academic career. Up until this point in time, I had found research a very solitary experience. The culture of the DReaM Group changed this. The group had, and continues to provide, a vibrant and nurturing environment in which to develop and test new ideas—however "half-baked" they may seem to begin with! I was also lucky in joining the group just as the rippling technique and proof planning were starting to make an impact. Through the group, and Alan's guidance, I have developed the career that I thought was beyond me. I will always be thankful and proud to have been part of the DReaM Group.

Gudmund I first experienced the DReaM Group as a PhD student at Heriot-Watt under the supervision of Andrew and Greg [Michaelson], then as a post-doc working with Alan [Bundy] in Edinburgh, and finally, as a lecturer at Heriot-Watt. My PhD, entitled "*Reasoning about Correctness Properties of a Coordination Language*", was my first real exposure to a research environment. At that point, I probably did not appreciate the emphasis of the group on sharing ideas (no matter how silly and undeveloped they may be) and the strong mantra of supporting and developing the careers of young researchers. This is something I have come to appreciate more and more over the years and I will always be indebted to Alan, Andrew and Greg for where I am today. They, and the DReaM Group as a whole, have played a major part in forming me as a researcher and I will always feel a part of the group wherever I may be in the future.

Teresa I joined the DReaM Group while I was doing my PhD (entitled "*Invariant Discovery and Refinement Plans for Formal Modelling in Event-B*") under the supervision of Andrew at Heriot-Watt University. Doing research was a very new and challenging experience for me, and initially I found it very nerve-racking to be part of a group whose members were so incredibly brilliant. However, I found the environment in the group to be very welcoming for young and inexperienced researchers. I also found the support given by the group through the EPSRC

Platform Grants to be a unique factor. It provided some sense of security in the world of academia that, as a young researcher, is very uncertain. The Platform Grants managed by Alan, and then by Andrew, gave me the possibility to continue my career in academia and led me in the path of pursuing my current research interests on computational creativity along with Simon Colton. I am very grateful to have been part of the DReaM Group, in particular for having been so lucky to be working with Andrew, who has been not only a guide for me in my work, but has also been a role model at the personal level; and Simon, who has led me to discover new exciting and innovative research paths.

References

1. J-R. Abrial, M. Butler, S. Hallerstede, and L. Voisin. An open extensible tool environment for Event-B. In *ICFEM*, LNCS **4260**, pages 588–605. Springer, 2006.
2. J-R. Abrial. *Modelling in Event-B: System and Software Engineering*. Cambridge University Press, 2010.
3. A. Bundy, D. Basin, D. Hutter, and A. Ireland. *Rippling: Meta-level Guidance for Mathematical Reasoning*. Cambridge University Press, 2005.
4. M. Butler and D. Yadav. An incremental development of the Mondex System in Event-B. *Formal Aspects of Computing*, 20(1), 2008.
5. M. Butler. Decomposition structures for Event-B. In *iFM*, LNCS **5423**, 2009.
6. S. Colton. *Automated Theory Formation in Pure Mathematics*. Springer, 2002.
7. S. Colton and S. Muggleton. Mathematical applications of Inductive Logic Programming. *Machine Learning*, 64:25–64, 2006.
8. K. Damchoom. *An Incremental Refinement Approach to a Development of a Flash-Based File System in Event-B*. PhD thesis, University of Southampton, 2010.
9. A.S Fathabadi and M. Butler. Applying Event-B atomicity decomposition to a multi-media protocol. In *FMCO*, LNCS **6286**, 2010.
10. A.S. Fathabadi, A. Rezazadeh, and M. Butler. Applying atomicity and model decomposition to a space craft system in Event-B. In *NFM*, LNCS **6617**, 2011.
11. G. Grov, A. Ireland, M.T. Llano, P. Kovacs, S. Colton, and J. Gow. Semi-Automated Design Space Exploration for Formal Modelling. arXiv:1603.00636.
12. G. Grov, A. Ireland, and M.T. Llano. Refinement plans for informed formal design. In *International Conference on Abstract State Machines, Alloy, B, VDM, and Z*, pages 208–222. Springer, 2012.
13. G. Grov, A. Ireland, M.T. Llano, P. Kovacs, S. Colton, and J. Gow. Semi-automated design space exploration for formal modelling. In *International Conference on Abstract State Machines, Alloy, B, TLA, VDM, and Z*, pages 282–289. Springer, 2016.
14. G. Grov, F. Mancini, and E. Mestl. Challenges for Risk and Security Modelling in Enterprise Architecture. In *The Practice of Enterprise Modeling*, pages 215–225, Cham, 2019. Springer.
15. S. Hallerstede, M. Leuschel, and D. Plagge. Refinement-Animation for Event-B - Towards a method of validation. In *ABZ*, volume 5977 of *LNCS*, pages 287–301. Springer, 2010.
16. A. Ireland. The use of planning critics in mechanizing inductive proofs. In A. Voronkov, editor, *International Conference on Logic Programming and Automated Reasoning (LPAR'92), St. Petersburg*, Lecture Notes in Artificial Intelligence No. 624, pages 178–189. Springer-Verlag, 1992. Also available from Edinburgh as DAI Research Paper 592.
17. A. Ireland and A. Bundy. Productive use of failure in inductive proof. 16(1–2):79–111, 1996. Also available as DAI Research Paper No 716, Dept. of Artificial Intelligence, Edinburgh.

18. A. Ireland and A. Bundy. Automatic verification of functions with accumulating parameters. *Journal of Functional Programming: Special Issue on Theorem Proving & Functional Programming*, 9(2):225–245, March 1999. A longer version is available from Dept. of Computing and Electrical Engineering, Heriot-Watt University, Research Memo RM/97/11.
19. A. Ireland, G. Grov, M.T. Llano, and M. Butler. Reasoned modelling critics: turning failed proofs into modelling guidance. *Science of Computer Programming*, 78(3), 2013.
20. A. Ireland and J. Stark. Proof planning for strategy development. *Annals of Mathematics and Artificial Intelligence*, 29(1-4):65–97, February 2001. An earlier version is available as Research Memo RM/00/3, Dept. of Computing and Electrical Engineering, Heriot-Watt University.
21. A. Ireland, G. Grov, and M. Butler. Reasoned modelling critics: turning failed proofs into modelling guidance. In *International Conference on Abstract State Machines, Alloy, B and Z*, pages 189–202. Springer, 2010.
22. A. Ireland, M.T. Llano, and S. Colton. The use of automated theory formation in support of hazard analysis. In Aaron Dutle, César A. Muñoz, and Anthony Narkawicz, editors, *NASA Formal Methods - 10th International Symposium, NFM 2018, Newport News, VA, USA, April 17-19, 2018, Proceedings*, volume 10811 of *Lecture Notes in Computer Science*, pages 237–243. Springer, 2018.
23. M.A. Jackson. *Problem Frames: Analysing and Structuring Software Development Problems*. ACM Press books. Addison-Wesley/ACM Press, 2001.
24. M. Johansson, L. Dixon, and A. Bundy. Conjecture synthesis for inductive theories. *J. Automated Reasoning*, 47:251–289, 10 2011.
25. P. Kovacs. Automating abstractions in formal modelling. BSc honour's thesis, Heriot-Watt University, 2015.
26. K.R.M. Leino. Dafny: An automatic program verifier for functional correctness. In *International Conference on Logic for Programming Artificial Intelligence and Reasoning*, pages 348–370. Springer, 2010.
27. Y. Lin. *The use of rippling to automate Event-B invariant preservation proofs*. PhD thesis, University of Edinburgh (School of Informatics), 2015.
28. Y. Lin, A. Bundy, G. Grov, and E. Maclean. Automating Event-B invariant proofs by rippling and proof patching. *Formal Aspects of Computing*, 31(1):95–129, 2019.
29. M.T. Llano, A. Ireland, and A. Pease. Discovery of invariants through automated theory formation. *Formal Aspects of Computing*, 26, 2011.
30. E. Maclean, A. Ireland, and G. Grov. Proof automation for functional correctness in separation logic. *Journal of Logic and Computation*, 2014.
31. R. Monroy, A. Bundy, and A. Ireland. Proof Plans for the Correction of False Conjectures. In F. Pfenning, editor, *5th International Conference on Logic Programming and Automated Reasoning, LPAR'94*, Lecture Notes in Artificial Intelligence, v. 822, pages 54–68, Kiev, Ukraine, 1994. Springer-Verlag. Also available from Edinburgh as DAI Research Paper 681.
32. M.T. Llano. *Invariant discovery and refinement plans for formal modelling in Event-B*. PhD thesis, Heriot-Watt University, UK, 2013.

Chapter 7
Human-Like Computational Reasoning: Diagrams and Other Representations

Mateja Jamnik

Abstract In this chapter, I give a personal account of my experience in Alan Bundy's DReaM group in the Department of Artificial Intelligence at the University of Edinburgh between the years of 1995 and 1998. Of course, the impact of this experience has been profound and long-lasting to this day. The culture and the nature of research work, the collaborations, the interests and the connections have endured, evolved and multiplied throughout this time. My own work in the DReaM group started by investigating human "informal" reasoning and formalising it in a diagrammatic theorem prover. After leaving Edinburgh, this work naturally evolved into combining diagrams with other representations in a uniform framework, as well as applying visual representations in other domains, such as reasoning with ontologies. But one of the fundamental questions remained unanswered, namely, how do we choose the right representation of a problem and for a particular user in the first place?

7.1 The DReaM Research Environment

Few factors influence a researcher's ethos regarding their work more than where and with whom they did their PhD project. I arrived to Alan Bundy's DReaM research group in the autumn of 1995, fresh from finishing a post-graduate Diploma in Computer Science at Cambridge. This was not exactly planned: I actually applied to do a PhD in the Cognitive Science Department at the University of Edinburgh. I was interested in humans, not machines. But given that I was a mathematician by my undergraduate degree and that I just finished a post-graduate degree in Computer Science, my application made it to Alan Bundy in the Department of Artificial Intelligence. I am so glad for this serendipity because the privilege has been immeasurable.

M. Jamnik (✉)
Department of Computer Science and Technology, University of Cambridge, Cambridge, United Kingdom
e-mail: mateja.jamnik@cl.cam.ac.uk

G. Michaelson (eds.), *Mathematical Reasoning: The History and Impact of the DReaM Group*, https://doi.org/10.1007/978-3-030-77879-8_7

The DReaM group at that time was a large and thriving community across multiple sites, covering the Universities of Edinburgh, Heriot-Watt and Napier, of very diverse people working on equally diverse research topics. The academic staff were Alan Bundy, Alan Smaill, Andrew Ireland and Helen Lowe.[1] They were working with numerous post-doctoral researchers including Ian Green, Toby Walsh, Richard Boulton, Julian Richardson and Geraint Wiggins. During my time in the group, I was part of a cohort of PhD students including Louise Dennis, Jon Whittle, Raul Monroy, Simon Colton, Francisco Cantu, Ian Frank, Jeremy Gow, Stephen Creswell and Jim Molony. We also collaborated internationally with scientists like Fausto Giunchiglia, Alessandro Armando and their groups in Italy, and Jörg Siekmann, Erica Melis, Dieter Hutter and their groups in Germany. Of course, the DReaM's ethos of creativity of thought and rigour of methodology have since spread around the world as we have pursued our careers across the globe and inevitably passed these values and skills to the next generation of researchers.

I could perhaps describe the diverse topics in mathematical reasoning that the DReaM members pursued as either formalising symbolic reasoning or formalising human-like reasoning.[2] Symbolic reasoning directions included proof planning, rippling, the systems Clam and λClam, induction, co-induction and hardware verification (Chaps. 1–4). Human-like reasoning directions included analogy, diagrammatic reasoning, ontologies and concept formation (Chaps. 5–8). Inevitably, this list is only partial, and all the chapters of this volume hopefully fill some of the gaps. I was particularly interested in the kinds of human-like reasoning that we could perhaps call "intuitive", or the kind that is inherently human and that is quite different to machine-oriented reasoning. Examples include the use of analogy, symmetry and diagrams.

Our daily lives as researchers were enriched by the visits of numerous scientists who shared their expertise and thoughts with us. Three visitors strongly shaped the direction that I took in my PhD research. Erica Melis from Saarbrücken was working on analogy reasoning at the time [23]. I was intrigued at how one can use examples of solutions in one problem to inspire and help us find a solution to a related problem. Erica mechanised this process in the context of proof. Whilst I did not use her work directly in my PhD, it turns out that my first project after my PhD was to mechanise learning of proof methods by analogy [17]. The second most memorable visitor was Predrag Janičič from Belgrade. He was interested in geometrical reasoning [18], which very much coincided with my interest in human visual reasoning. Predrag also became a close friend, and I could speak his language, so we had our own way of communicating. Finally, perhaps the most influential visitor in the DReaM group for me was Alan Robinson. He came to Edinburgh early

[1] I apologise if my poor memory is not serving me well and I mixed people up or inevitably forgot to mention some.

[2] This divide is perhaps a little artificial since all of our work was motivated by the goals of artificial intelligence, namely we were trying to computationally model human mathematical reasoning. Alan Bundy's Chap. 1 of this volume gives a more precise overall description of our work.

on in my PhD and was interested, like me, in “intuitive” or “informal” reasoning. Our discussions surrounded the distinction between a visual or spatial representation and the more usual machine symbolic representation. Alan Robinson showed me a number of “proofs without words” that he encountered, and some of them became my toy or working examples to represent, solve and mechanise the process. He believed that our ability to “see” the truthfulness of a statement is one of the really fundamental components of the human mathematical cognitive repertoire. Later, when my PhD research was published in a book [15], he kindly wrote in the foreword that in this work I *“found an explanation of at least part of the mystery of how humans are able to ‘see’ the truth of certain mathematical propositions merely by contemplating appropriate diagrams and constructions”*.

What struck me most about the DReaM group was his openness to discuss any topic that anybody was interested in. Clearly, Alan Bundy nurtured a kind and supportive research environment in which everyone could do their best work. His intellectual generosity was boundless. The Blue Book Notes (see Chap. 1) provided the opportunity for sharing and discussing our research with the group. They laid the ground for exploring novel and half-baked ideas that most often developed into mature and original scientific contributions. Alan led the DReaM group in an organised way that taught me how to be a supervisor and a mentor to my own students and post-docs. This is perhaps best demonstrated by numerous “How-to...” guides that Alan wrote, for example, “How to be my student”, “How to write an informatics paper”, “Writing a good grant proposal”, “The Researchers Bible” and “How to say no”.[3]

If I were to summarise the enduring influence that the DReaM group had on me, then I would put in the first place the intellectual generosity that I try to bestow on my own research group today. I learnt the importance of rigorous methodology and the place for heuristics to guide the automation of reasoning. Perhaps uniquely at that time, our work provided a human-oriented perspective on artificial intelligence that remained the main motivation for my research. Finally, my time in Alan’s DReaM group instilled in me the importance of an interdisciplinary and collaborative approach to research, which I think is key to innovation in AI today.

7.2 Diagrammatic Reasoning

Despite the fact that diagrams have been used in mathematics since the time of Aristotle and Euclid, the invention of formal axiomatic logic at the end of the nineteenth century in the sense of Frege, Russell and Hilbert denied diagrams a *formal* role in theorem proving. Diagrams were only used informally for illustrating a formal proof and for suggesting proof steps but were formally superfluous.

[3]Many of these can be found on Alan Bundy’s web page: https://sweb.inf.ed.ac.uk/bundy/.

Fortunately, the end of the twentieth century started to see a redressing of this issue [2, 5]. Examples include formalised logical systems of diagrams [10, 13, 35]. This directly abolished the widely held Hilbertian theoretical objections to diagrams being used in proofs. Our work on Diamond was amongst these: it pioneered the construction of purely diagrammatic proofs where diagrams and their manipulations are the proof [15, 16].[4] The motivation for this work was rooted in formalising some of the "informal" reasoning that humans do in mathematics when using diagrams.

Take for example, the diagram in Fig. 7.1. It takes only secondary school-level knowledge of mathematics to understand that the diagram is about the sum of odd natural numbers. We can "see" that the theorem is true not only for the example in the diagram of $n = 6$, but for any value of n. In other words, the simple procedure of splitting a square into the so-called ells works in general. Diamond tackles this problem, in addition to a number of other, the so-called proofs without words, many of which can be found in Nelsen's books [25, 26] and Gardner's mathematical recreations [8, 9].

Diamond's theorems are in the domain of algebraic mathematics about natural numbers that can be expressed as diagrams in a discrete space and are inductive over a parameter. But there is a problem, namely, such diagrams are concrete in nature, so abstractions such as ellipsis need to be used to express the general diagram (and proof) for all values of the parameter. These abstractions are difficult to keep track of whilest manipulating. So we proposed a solution: to use schematic proofs.

Schematic proofs are based on the mathematical notion of the ω-rule that says that for the natural numbers 0, 1, 2, . . .:

$1 + 3 + 5 + \ldots + (2n - 1) = n^2$

Fig. 7.1 The theorem is about the sum of the first n odd natural numbers. It represents the example of a case for $n = 6$. The proof starts from the RHS of the theorem n^2 and takes a square. Then, the square is split into a sequence of nesting and increasing in size, the so-called ells. Each ell represents a subsequent natural number: there are two edges, each of size n, but the joining vertex has been counted twice; hence an ell is $2n - 1$

[4]This work was done for my PhD with Alan Bundy and Ian Green as my supervisors.

$$\frac{\phi(0), \quad \phi(1), \quad \phi(2), \quad \dots}{\forall x.\phi(x)}.$$

That is, if we can prove $\phi(n)$ for $n = 0, 1, 2, \dots$, then we can infer that $\phi(x)$ for all natural numbers x. Clearly, the ω-rule is not very practical for automation, since it requires the proof of an infinite number of premises to prove its conclusion. A more practical alternative is the constructive ω-rule that has an additional condition: if all premises $\phi(n)$ can be proved in a *uniform* way, that is, there exists an effective procedure, $proof_{\phi}$, which takes a natural number n as input and returns a proof of $\phi(n)$ as output, then we can conclude the universal statement:[5]

$$proof_{\phi}(n) \vdash \phi(n).$$

One such effective procedure is, for example, a recursive program. Now $proof_{\phi}$ can be a recursive procedure that formalises our notion of schematic proof where the number of steps in the proof depends on the parameter n. We used this notion in formalising diagrammatic proofs in Diamond.[6]

Diamond's theorems are expressed as diagrams for some concrete values, that is, ground instantiations of a theorem. The initial diagram is manipulated using some geometric operations. The sequence of geometric operations on a diagram represents the inference steps of a diagrammatic proof. In the above example, the inference step is splitting an ell from a square to produce an ell and a smaller square. The set of all available operations defines the proof search space. Next, Diamond automatically extracts a general pattern from these proof instances and captures it in a recursive program that constitutes a general diagrammatic proof for the universally quantified theorem. The constructive ω-rule justifies the step from schematic proofs to theoremhood. In Diamond, the diagrammatic schematic proof is formalised as

$$\begin{aligned} proof_{\phi}(n+1) &= \mathscr{A}(n+1), proof_{\phi}(n) \\ proof_{\phi}(0) &= \mathscr{B}, \end{aligned}$$

where $\mathscr{A}(n+1)$ consists of a sequence of diagrammatic operations, and the number of applications of each operation is (linearly) dependent on n. $\mathscr{B}$ is a possibly empty basis, that is, no additional operation is required to complete the proof.

The generated program capturing the schematic proof still needs to be verified to be correct. This is something that human mathematicians often omit, and hence history of mathematics is full of erroneous proofs (see Cauchy's proof of Euler's theorem as reported by Lakatos in [21] and in Chap. 1 of this volume). The

[5]From the logical point of view, the constructive ω-rule (and also the ω-rule) is a stronger alternative to mathematical induction, where the generation of proofs for all instances is satisfied by the requirement for the effective procedure, such as a recursive function.

[6]The constructive ω-rule and schematic proofs have previously been implemented for arithmetic theorems and their symbolic proofs by another DReaM member, Siani Baker [3].

verification requires meta-level reasoning about the proof, rather than the object-level theorem, and is done by induction:

$$proof_{\phi}(0) \vdash \phi(0)$$
$$proof_{\phi}(n) \vdash \phi(n) \implies proof_{\phi}(n+1) \vdash \phi(n+1).$$

The work on automation of diagrammatic proofs in Diamond provides important information on proof procedure construction. It exposes the importance of representing diagrammatic expressions so that general reasoning techniques can be applied to them. Furthermore, it provides an insight into how diagrams and purely diagrammatic inferences can be used in formal proofs.

7.3 Heterogeneous Reasoning

Picking up any mathematical book reveals that many theorems are proved using symbolic inference steps as well as diagrams. We call these heterogeneous proofs: examples of two such proofs can be seen in Fig. 7.2. In the first example, a theorem about triangular numbers is proved by transforming it with symbolic inferences into an expression that then has a compelling diagrammatic proof.[7] In the second example, the theorem asserts a statement about a bitmap image that clearly requires the use of image processing steps to then combine them with symbolic inferences.

There exist tools for combining diverse systems (e.g., OpenBox [4], Omega [36], HETS [24]), but they do not allow mixing of representations. Indeed, most mechanised theorem provers use only symbolic representations, like different types of logic. Whilst Diamond (and other diagrammatic theorem provers like Speedith [42]) constructs proofs using only diagrammatic inference steps, not all theorems can be expressed with diagrams. Moreover, human mathematicians typically use not only multiple, but also informal representations such as natural language or images within the same problem for different parts of the solution.

We designed and built a heterogeneous reasoning framework MixR [41] where different existing symbolic as well as diagrammatic reasoners can be used at the same time so that symbolic and diagrammatic proof steps can be interleaved within the same proof.[8] Furthermore, when logical formalisation of a particular representation (e.g., images, natural language or audio) is not tractable, we can embed such data in existing provers and still enable informal heterogeneous reasoning with these opaque objects within an otherwise formal proof.

The MixR framework provides a generic infrastructure for extending existing general-purpose theorem provers with heterogeneous reasoning in the form of

[7] Notice that there is no compelling completely diagrammatic proof of the original expression of the theorem, thus the need to mix symbolic and diagrammatic inference steps.

[8] This work was done in collaboration with my PhD student Matej Urbas.

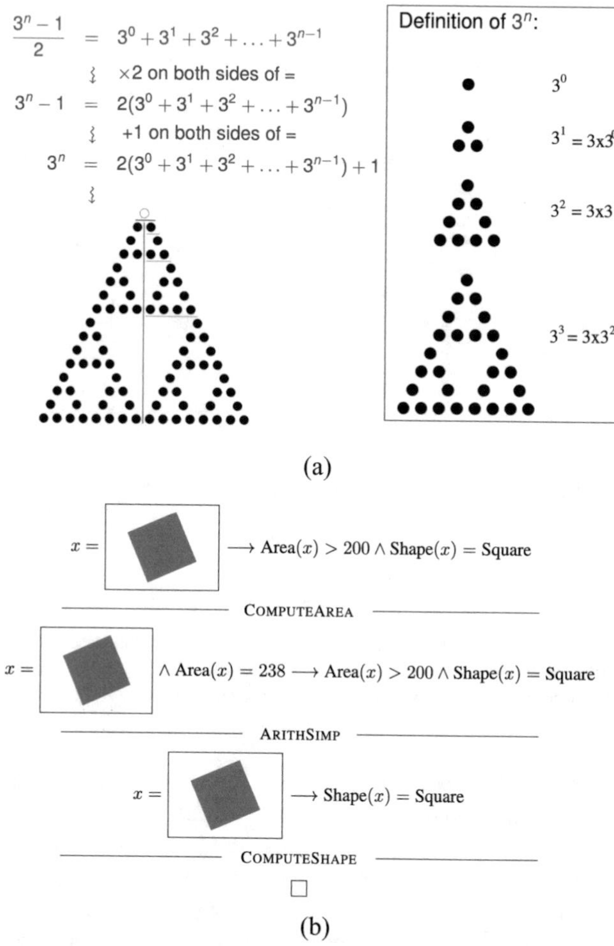

Fig. 7.2 Examples of two heterogeneous proofs: (**a**) a few symbolic steps are followed by transformation of the formula into a diagram followed by diagrammatic proof steps to prove a theorem about triangular numbers; (**b**) here the heterogeneous proof consists of three proof steps: the ComputeArea inference step is heterogeneous and takes a bitmap image and extracts some information (the area of the square) that is expressed in the symbolic language; the ArithSimp inference step is symbolic; the ComputeShape is also a heterogeneous inference step—it extracts that the bitmap shape is a square and thus resolves the implication

heterogeneous logic. The crucial part of our heterogeneous logic is the mechanism, called placeholders, which embeds foreign data into formulae of existing theorem provers so that it can be dealt with using external tools. This data is directly embedded into formulae of a prover that treats them as primitive objects that can be reasoned with its standard inference engine. When required, the reasoner can invoke external tools on this data to obtain new knowledge. Our approach using

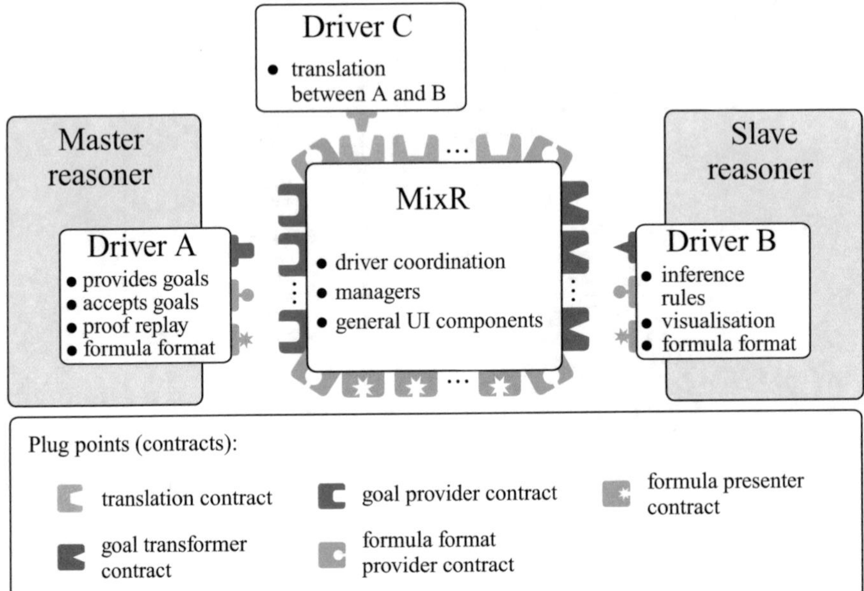

Fig. 7.3 MixR's architecture with hypothetical drivers. The central box represents MixR's core. It contains the implementation of heterogeneous logic components, general UI components and driver plug points. Drivers surround MixR's core and plug into it through the plug points

placeholders removes the need for translations between representations, which is particularly useful when no such translation is available or even possible (e.g., diagrammatic representations from CAD tools, images and signal processing).

MixR is an implementation of this heterogeneous logic and placeholders and enables the integration of arbitrary existing theorem provers of any modality with each other into new heterogeneous systems. A tool developer can plug their chosen reasoners into MixR by writing MixR drivers for them. MixR, in turn, integrates them with each other into a new heterogeneous reasoning system. For example, we plugged Speedith [42] for spider diagrams and Isabelle [28] for sentential higher-order logic into MixR to create the Diabelli [40] heterogeneous reasoning system. We also integrated image processing with symbolic reasoning into PicProc [41] that can prove a theorem in Fig. 7.2b. MixR provides a user interface as well as an application programming interface (API) for drivers. Using the API, the drivers can share, translate and visualise formulae of various modalities. They may also apply foreign inference steps and query other drivers to invoke foreign reasoning tools. The architecture of MixR is illustrated in Fig. 7.3.

Many reasoning tools, representations and visualisation aids in artificial intelligence exist mostly in isolation, specialised in their specific domains. Bringing them together in a simple, flexible and formal way, as in MixR, allows them to contribute to the problem-solving/theorem-proving tasks. This better models what people do in problem solving, it allows developers to easily design systems that are flexible according to the needs of the end users, and it enables us to take advantage of the existing powerful technology in a novel and sustainable way.

7.4 Accessible Diagrammatic Reasoning About Ontologies

One of our main motivating factors for computationally modelling reasoning with diagrams has been the fact that people use them and find them intuitive and accessible. The barrier to entry for explaining problems and their solutions is lower using diagrams than symbolic logical formalisms. One domain that routinely requires some level of formal reasoning but involves a range of different stakeholders is ontologies. Ontologies are a common knowledge representation paradigm, but they frequently have accessibility issues due to unfamiliarity of domain experts with symbolic notations (e.g., DL, OWL). Some visualisation facilities have been implemented [14, 22], but their focus is expressing and editing ontologies rather than reasoning with and about ontologies.

Ontologies represent knowledge in a domain with definitions of concepts, their properties and relations between concepts. Reasoning with ontologies is done with a justification algorithm [19] that selects a minimal set of axioms responsible for entailment. There is empirical evidence [12] that confirms that stakeholders find it difficult to get from the justification to the explanation of the reasons for the particular selected axioms entailing the problem. Thus, a number of symbolic theorem provers have been implemented, which construct a symbolic explanation for justification–entailment pair. Unfortunately, these proofs have the same inaccessibility issues as before: domain experts are not familiar with their symbolic notations.

In order to address the inaccessibility of symbolic notations, we devised a visual theorem prover, *iCon*, that uses a visual language to represent and reason with ontologies.[9] The input to iCon is a justification–entailment pair expressed as diagrammatic axioms (justifications) and a diagrammatic theorem (ontology entailment). The output is an interactively constructed proof using applications of diagrammatic inference rules that explains how the entailment follows from the axioms.

The visual language of iCon, concept diagrams [37], covers almost all of the standard ontology language OWL 2. Empirical studies demonstrate the accessibility of concept diagrams compared to competing diagrammatic and symbolic notations [33]. Concept diagrams consist of curves (circles, as in Euler and Venn diagrams) that represent ontology classes (they are sets), dots and spiders that

---

[9]This work was done during the Leverhulme Trust funded project "ARD: Accessible Reasoning with Diagrams" in collaboration with Gem Stapleton, Zohreh Shams, Yuri Sato, Sean Mcgrath and Andrew Blake.

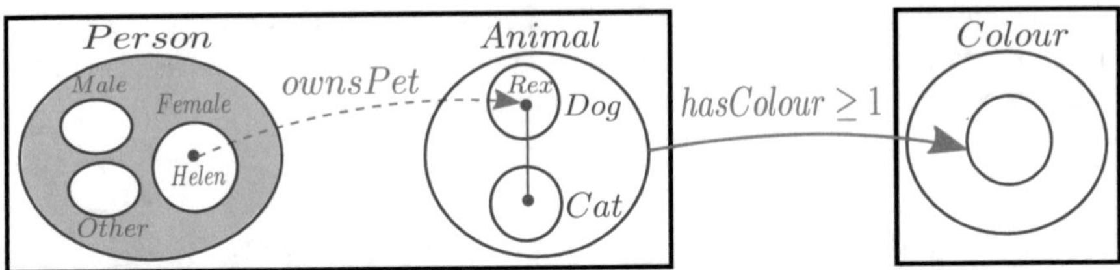

Fig. 7.4 Example of a concept diagram

represent individuals in classes, and arrows that represent object properties. There are also boundary rectangles to denote all individuals in the world, and shading to place an upper bound on the cardinality of the sets. Complete formalisation of concept diagrams is given in [38].

Figure 7.4 shows a concept diagram that has 2 bounding rectangles. Spatial relationships between parts of the diagram convey information, for example, that *Person* and *Animal* represent disjoint sets, since the two corresponding curves are disjoint. We can also see that *Helen* is a *Female* person, due to the location of the (red) dot labelled *Helen*. A dot connected by a line to another dot is called a *spider*, and it signals that it is not clear which set an individual belongs to. For example, in Fig. 7.4, *Rex* could be either a *Cat* or a *Dog*. The region outside of *Other*, *Male* and *Female* is shaded, which means that there is no person who is neither a *Female*, a *Male* nor *Other*. The dashed arrow *ownsPet* connects the dot *Helen* to *Rex*. This means that *Helen* owns *Rex* as her pet, but she can own pets of other types too. Unlike dashed arrows, solid arrows mean that the source is related to *only* the target. So, the colours that an animal can have cannot be outside the set *Colour*. Together with the arrow annotation ≥ 1, this means that all animals have *at least* one colour.

iCon consists of an inference engine and the graphical user interface. The inference engine contains a collection of inference rules, applies inference rules to diagrams and manages proofs. The inference rules can be either symbolic (conjunction elimination or identity) or diagrammatic. The diagrammatic inference rules come from the ontology community's standard set of inference rules for OWL 2 RL [27], introduced by the W3C in [43]. In order to construct a proof for a justification–entailment pair, we equipped iCon's inference engine with diagrammatic versions of the symbolic inference rules for OWL 2 RL. Diagrammatic inference rules rewrite the diagrams representing the premises of a proof state in order to make them identical to the goal of the proof state. In contrast to a symbolic proof, which is typically inaccessible to domain experts, this results in a diagrammatic proof, which is empirically evidenced to be more accessible [1, 33]. Figure 7.5 illustrates an example of such a diagrammatic inference rule. Reasoning in ontologies most commonly involves entailments, that is, checking if the set of axioms is consistent, coherent or for query answering. Thus, proofs will often be about finding out why a set of axioms is inconsistent or incoherent so that the ontology can be repaired. An example of both a symbolic and iCon's diagrammatic proofs of a theorem about inconsistency can be seen in [34].

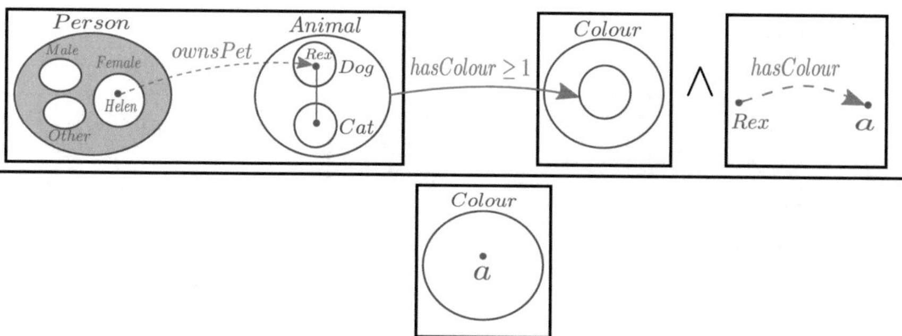

Fig. 7.5 Example of iCon's diagrammatic inference rule

Ontologies are frequently used in the real world by diverse stakeholders, so it is paramount to make working with them accessible. Current symbolic reasoners for ontologies provide only a minimal set of axioms for entailments without explanations for these entailments or indeed lack of entailment. In contrast, iCon's diagrammatic proof provides not only an explanation for the entailment that exposes the interaction between the minimal set of axioms, but also an accessible evidence and clues for how to repair the ontology when it is found inconsistent or incoherent. Thus, iCon can be effectively used for reasoning about and debugging of ontologies.

7.5 How to Choose a Representation

So far, we showed how diagrams can be used for formal reasoning, how architectures can be built to enable reasoning with diverse types of representations and indeed tools, and how we can formally reason with diagrams about ontologies. But the question remains: given a problem that we want to solve, how do we choose the representation that is best suited for solving it and that is most appropriate for the user who is trying to solve it? Cognitive science has firmly established that choosing an effective representation can yield dramatic improvements in human problem-solving performance [7, 20] and substantially enhance learning [6]. This is what we are currently investigating in an interdisciplinary project on human-like computing, which has Alan Bundy as one of its advisers.[10] We are combining artificial intelligence, mathematics and cognitive science to investigate human cognitive abilities to find representations that suitably match problems, and the process by which humans adapt or switch between representations. We are devising a foundational theory and building computational models of the critical role that

[10]This work started during the EPSRC funded projects "How to (Re)represent it?" and "Automating Representation Choice for AI Tools" in collaboration with Peter Cheng, Daniel Raggi (also an ex-DReaMer), Grecia Garcia Garcia, Aaron Stockdill, Holly Sutherland and Gem Stapleton.

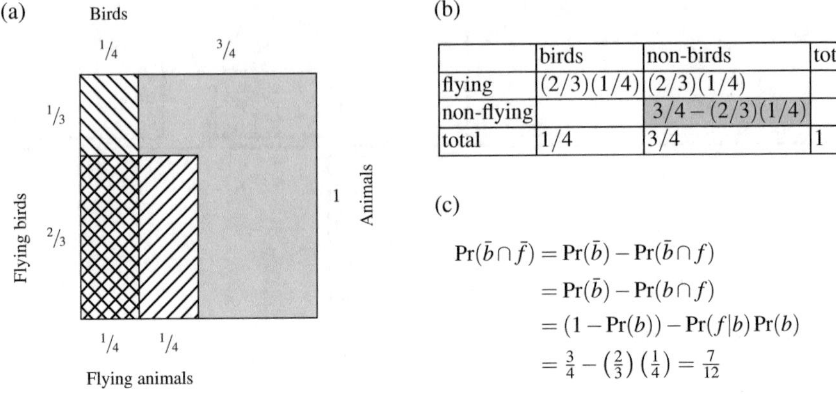

	birds	non-birds	total
flying	(2/3)(1/4)	(2/3)(1/4)	
non-flying		3/4 − (2/3)(1/4)	
total	1/4	3/4	1

$$
\begin{aligned}
\Pr(\bar{b} \cap \bar{f}) &= \Pr(\bar{b}) - \Pr(\bar{b} \cap f) \\
&= \Pr(\bar{b}) - \Pr(b \cap f) \\
&= (1 - \Pr(b)) - \Pr(f|b)\Pr(b) \\
&= \tfrac{3}{4} - \left(\tfrac{2}{3}\right)\left(\tfrac{1}{4}\right) = \tfrac{7}{12}
\end{aligned}
$$

Fig. 7.6 Bird probability example. (**a**) Geometric representation—the solution is the area of the solid shaded region $\frac{3}{4} - \left(\frac{2}{3}\right)\left(\frac{1}{4}\right) = \frac{7}{12}$. (**b**) Contingency table representation—the solution is in the shaded cell. (**c**) Bayesian representation

representations play in problem solving, and automating them in a new generation of adaptive AI systems [30–32, 39].

To illustrate our approach, consider this problem in probability:

One quarter of all animals are birds. Two thirds of all birds can fly. Half of all flying animals are birds. Birds have feathers. If X is an animal, what is the probability that it's not a bird and it cannot fly?

Here are three different ways one can go about solving this (see Fig. 7.6):

1. You could divide areas of a rectangle to represent parts of the animal population that can fly and parts that are birds.
2. You could use contingency tables to enumerate in its cells all possible divisions of animals with relation to being birds or being able to fly.
3. You could use formal Bayesian notation about conditional probability.

Which of these is the most effective representation for the problem? It depends; the first is probably best for school children; the last for more advanced mathematicians. How can this choice of appropriate representation be mechanised? We are interested to find out:

- What are the formal mathematical and cognitive foundations for choosing an effective representation of a problem?
- Can we develop new cognitive theories that allow us to understand the relative benefits of different representations of problems and their solutions, including taking into account individual differences?
- How can we automate an appropriate choice of problem representation for both humans, taking into account individual differences, and machines to improve human–machine communication?

- Can we build an AI tutoring system, aimed at mathematical problems, that incorporates personalised representation choices and improves users' abilities to solve problems?

We distinguish between *cognitive* and *formal* properties of a representation, in an approach that radically, but systematically, reconfigures previously descriptive accounts of the nature of representations [11]. We use this to devise methods for measuring competency in alternative representation use and also to engineer a system to automatically select representations. *Cognitive properties* characterise cognitive processes demanded of a particular representation (e.g., problem state space characteristics; applicable state space search methods; attention demands of recognition; inference operator complexity [6]). *Formal properties* characterise the nature of the content of the representation domain (e.g., operation types like associative or commutative, symmetries, coordinate systems, quantity or measurement scales).

We devised a novel encoding for taxonomising formal and cognitive properties of problems and representational systems [30, 32]. We catalogue formal properties using templates of attributes that (currently) the developer of the system assigns values to. The attributes encode the informational content of the question and a representational system. Table 7.1 gives snippets from a formal property catalogue for the above *Birds* problem stated in the natural language representation. The colours code the importance of the property relative to the information content (top to bottom in decreasing importance). Table 7.2 gives snippets of the catalogue of formal properties for the Bayesian representational system (used in the solution in Fig. 7.6c). Any representational system and problem expressed in it can be encoded using this description language.

We built algorithms that automatically analyse these encodings for a given problem (like the one in Table 7.1) with respect to candidate representational systems (like the one in Table 7.2) in order to rank the representations, and ultimately suggest the most appropriate one. This analysis is largely based on *correspondences* between the properties of representational systems and their relative importance for a given problem. For example, the correspondences between the natural language formulation of the example and the Bayesian one are translational/morphism-like pairs, such as `ratio` $\rightsquigarrow$ `real`, `given` $\rightsquigarrow$ $|$, `probability` $\rightsquigarrow$ Pr and `intersection` $\rightsquigarrow$ $\cap$.

Similarly to formal properties, we devised a catalogue of 9 critical cognitive properties. They span spatial and temporal scales (icons to whole displays and seconds to tens of minutes), numerous cognitive processes and the mapping of information between symbols/expressions and concepts. The attributes of cognitive properties characterise the cognitive cost, that is, the difficulty of using that representational system for problem solving. We designed weighting functions to compute overall values of the cognitive cost for each property: they are based on a problem at hand, a typical user and utilise the taxonomy of formal properties.

To adjust cognitive costs from a typical user to individual's abilities, we devised a small but diverse set of user profiling tests. The measures extracted from these

Table 7.1 Formal properties of the *Birds* problem in its natural language representation (note colour)

Kind	Value
Error allowed	0
Answer type	Ratio
Primitives	Probability, and, not
Types	Ratio, class
Patterns	_:ratio of _:class are _:class, probability of _:class and _:class
Facts	Bayes' theorem, law of total probability, unit measure, additive inverse, ...
Tactics	Deduce, calculate
Primitives	One, quarter, all, animals, birds, two, thirds, can, fly, half, flying, X, animal, probability, cannot
Related primitives	Times, divided_by, plus, minus, equals, union, intersection, probability, zero, ...
# of primitives	67
# of distinct primitives	31
# of statements	5
Primitives	Feathers
Related primitives	Beast, animate, creature, wing, aviate, flock, fowl, dame, carnal, being, fauna, ...

Table 7.2 A section of formal properties for Bayesian representational system

Kind	Value
Types	Real, event
Primitives	$\Omega, \emptyset, 0, 1, =, +, -, *, \div, \cup, \cap, \backslash, \bar{}$, Pr, |
g-complexity	Type-2
Facts	Bayes' theorem, law of total probability, non-negative probability, unit measure, sigma additivity, commutativity ...
Tactics	Rewrite, arithmetic calculation
i-complexity	3
Rigorous	TRUE

profiles enable us to scale the level of contributions of each cognitive property to the overall cost of a representational system for an individual. We operationalised the encoding of cognitive properties by automating heuristics that encode user preferences and level of expertise to influence the ranking of potential candidate representational systems.

In this chapter, we are laying the foundations for understanding formal and cognitive properties that affect the choice of representations in problems solving. Our prototype implementations of the algorithms that carry out this analysis show that it is possible to model such processes computationally. We are now applying these foundations in applications such as personalised AI tutoring systems.

7.6 Future Directions

The overarching theme of the work reported here, and common to many past and present DReaM group members, is about computationally modelling human reasoning. The enduring legacy of the DReaM group and our common interests mean that in a number of these projects we continue with existing and establish new collaborations with the past and present DReaMers. For example, Alan Bundy is serving on the advisory board of my project about representation choice and AI tools, and Alison Pease is helping us with her HRL system [29] in our mathematical education project.

The aim of my work is to make AI systems more human-like in the way they interact with users, in the representations that they choose for this interaction, in the methods that they employ to solve problems and in the explanations that they provide alongside their solutions. There are many future directions, especially with respect to fully automating some of these processes and scaling them up to general real-world AI systems. In particular, we are currently developing automated methods for a diagrammatic reasoning system to *discover* new, intuitive solutions to mathematical problems. We are also investigating how we can make theorem provers construct proofs with methods at a level of abstraction and with a level of automation that human mathematicians find appealing. Furthermore, we are marrying statistical with symbolic and knowledge-based approaches to machine learning in order to enhance machine-oriented with human-oriented inference. The results are AI systems that produce solutions from fewer examples and with better explanations of the solutions. There are many applications of this work, but we are focusing on education and developing a new generation of adaptive AI tutoring systems, and on medicine and building integrative data models for clinical decision support systems in personalised cancer medicine.

There is currently much excitement about artificial intelligence and its impact on society. Most of the work that is generating this excitement is due to impressive results of statistical machine learning. However, these machine-oriented methods produce solutions that often lack explanations and use representations that are inaccessible to humans. My research is motivated by human reasoning, so I employ symbolic learning and knowledge-based reasoning as well as diverse representations to enhance this learning and inference. Interdisciplinarity and collaboration have always been at the centre of the DReaM group research ethos, and they have therefore undoubtedly shaped me and my work. Both are key to advancing the field and building a new generation of AI systems that are transparent and have a good cognitive model of the user to be adaptable and to produce explanations understandable to humans.

Acknowledgments I am thankful to all of my collaborators in the work reported here, including Alan Bundy, Ian Green, Matej Urbas, Gem Stapleton, Zohreh Shams, Yuri Sato, Sean Mcgrath, Andrew Blake, Peter Cheng, Daniel Raggi, Aaron Stockdill, Grecia Garcia Garcia and Holly Sutherland.

References

1. Alharbi, E., Howse, J., Stapleton, G., Hamie, A., Touloumis, A.: Visual logics help people: An evaluation of diagrammatic, textual and symbolic notations. In: IEEE Symposium on Visual Languages and Human-Centric Computing, pp. 255–259. IEEE (2017)
2. Anderson, M., Meyer, B., Oliver, P. (eds.): Diagrammatic Representation and Reasoning. Springer (2001)
3. Baker, S., Smaill, A.: A proof environment for arithmetic with the omega rule. In: J. Calmet, J. Campbell (eds.) Integrating Symbolic Mathematical Computation and Artificial Intelligence, no. 958 in Lecture Notes in Computer Science, pp. 115–130. Springer (1995)
4. Barker-Plummer, D., Etchemendy, J., Liu, A., Murray, M., Swoboda, N.: Openproof: A flexible framework for heterogeneous reasoning. In: G. Stapleton, J. Howse, J. Lee (eds.) Diagrams, *Lecture Notes in Artificial Intelligence*, vol. 5223, pp. 347–349. Springer (2008)
5. Chandrasekaran, B., Glasgow, J., Narayanan, N. (eds.): Diagrammatic Reasoning: Cognitive and Computational Perspectives. AAAI Press/MIT Press, Cambridge, MA (1995)
6. Cheng, P.: Electrifying diagrams for learning: principles for effective representational systems. Cognitive Science **26**(6), 685–736 (2002)
7. Cheng, P., Lowe, R., Scaife, M.: Cognitive science approaches to diagrammatic representations. Artificial Intelligence Review **15**(1-2), 79–94 (2001)
8. Gardner, M.: Mathematical Circus. Vintage, New York (1981)
9. Gardner, M.: Knotted Doughnuts and Other Mathematical Entertainments. W.H. Freeman and Company, New York (1986)
10. Hammer, E.: Logic and visual information. CSLI Press, Stanford, CA (1995)
11. Hegarty, M.: The cognitive science of visual-spatial displays: Implications for design. Topics in Cognitive Science **3**, 446–474 (2011)
12. Horridge, M., Parsia, B., Sattler, U.: Lemmas for justifications in OWL. In: 22nd International Workshop on Description Logics, vol. 477. CEUR-WS.org (2009)
13. Howse, J., Stapleton, G., Taylor, J.: Spider Diagrams. LMS JCM **8**, 145–194 (2005)
14. Itzik, N., Reinhartz-Berger, I.: SOVA - A tool for semantic and ontological variability analysis. In: Joint Proceedings of the CAiSE 2014 Forum and CAiSE 2014 Doctoral Consortium, vol. 1164, pp. 177–184. CEUR-WS.org (2014)
15. Jamnik, M.: Mathematical Reasoning with Diagrams: From Intuition to Automation. CSLI Press, Stanford, CA (2001)
16. Jamnik, M., Bundy, A., Green, I.: On automating diagrammatic proofs of arithmetic arguments. Journal of Logic, Language and Information **8**(3), 297–321 (1999)
17. Jamnik, M., Kerber, M., Pollet, M.: Automatic learning in proof planning. In: F. van Harmelen (ed.) Proceedings of 15th ECAI, pp. 282–286. European Conference on Artificial Intelligence, IOS Press (2002)
18. Janicic, P.: GCLC - A tool for constructive Euclidean geometry and more than that. In: A. Iglesias, N. Takayama (eds.) Mathematical Software - ICMS, *Lecture Notes in Computer Science*, vol. 4151, pp. 58–73. Springer (2006)
19. Kalyanpur, A.: Debugging and repair of owl ontologies. Ph.D. thesis, The University of Maryland (2006)
20. Kotovsky, K., Hayes, J.R., Simon, H.A.: Why are some problems hard? Cognitive Psychology **17**, 248–294 (1985)
21. Lakatos, I.: Proofs and Refutations: The Logic of Mathematical Discovery. Cambridge University Press, Cambridge, UK (1976)
22. Lohmann, S., Negru, S., Haag, F., Ertl, T.: Visualizing ontologies with VOWL. Semantic Web **7**(4), 399–419 (2016)
23. Melis, E.: A model of analogy-driven proof-plan construction. In: C. Mellish (ed.) Proceedings of the 14th IJCAI, pp. 182–189. International Joint Conference on Artificial Intelligence, Morgan Kaufmann, San Francisco, CA (1995)
24. Mossakowski, T., Maeder, C., Lüttich, K.: The Heterogeneous Tool Set, HETS. In: TACAS, *LNCS*, vol. 4424, pp. 519–522. Springer (2007)

25. Nelsen, R.: Proofs without Words: Exercises in Visual Thinking. Mathematical Association of America, Washington, DC (1993)
26. Nelsen, R.: Proofs without Words II: Exercises in Visual Thinking. Mathematical Association of America, Washington, DC (2001)
27. The OWL2 web ontology language. URL https://www.w3.org/TR/owl2-direct-semantics/. Retrieved Dec 2019
28. Paulson, L.: Isabelle: A generic theorem prover. No. 828 in Lecture Notes in Computer Science. Springer (1994)
29. Pease, A.: A computational model of Lakatos-style reasoning. Ph.D. thesis, Edinburgh University, UK (2007)
30. Raggi, D., Stapleton, G., Stockdill, A., Jamnik, M., Garcia Garcia, G., C.-H. Cheng, P.: How to (Re)represent it? In: 32th IEEE International Conference on Tools with Artificial Intelligence, pp. 1224–1232. IEEE (2020)
31. Raggi, D., Stockdill, A., Jamnik, M., Garcia Garcia, G., Sutherland, H., C.-H. Cheng, P.: Dissecting representations. In: A. Pietarinen, P. Chapman, L. Bosveld-de Smet, V. Giardino, J. Corter, S. Linker (eds.) Diagrams: Diagrammatic Representation and Inference, *LNCS*, vol. 12169, pp. 144–152. Springer (2020)
32. Raggi, D., Stockdill, A., Jamnik, M., Garcia Garcia, G., Sutherland, H., Cheng, P.: Inspection and selection of representations. In: C. Kaliszyk, E. Brady, A. Kohlhase, C. Sacerdoti-Coen (eds.) Intelligent Computer Mathematics (CICM), *Lecture Notes in Computer Science*, vol. 11617, pp. 227–242. Springer (2019)
33. Sato, Y., Stapleton, G., Jamnik, M., Shams, Z.: Human inference beyond syllogisms: an approach using external graphical representations. Cognitive Processing **20**(1), 103–115 (2019)
34. Shams, Z., Jamnik, M., Stapleton, G., Sato, Y.: iCon: A diagrammatic theorem prover for ontologies. In: F. Wolter, M. Thielscher, F. Toni (eds.) Principles of Knowledge Representation and Reasoning: Proceedings of the 16th International Conference, KR 2018, pp. 204–205. AAAI Press (2018)
35. Shin, S.: The Logical Status of Diagrams. Cambridge University Press, Cambridge, UK (1995)
36. Siekmann, J., Benzmüller, C., Brezhnev, V., Cheikhrouhou, L., Fiedler, A., Franke, A., Horacek, H., Kohlhase, M., Meier, A., Melis, E., Moschner, E., Normann, I., Pollet, M., Sorge, V., Ullrich, C., Wirth, C.P., Zimmer, J.: Proof development with Ω. In: A. Voronkov (ed.) 18th Conference on Automated Deduction, no. 2392 in Lecture Notes in Artificial Intelligence, pp. 144–149. Springer (2002)
37. Stapleton, G., Compton, M., Howse, J.: Visualizing OWL 2 using diagrams. In: IEEE Symposium on Visual Languages and Human-Centric Computing, pp. 245–253. IEEE (2017)
38. Stapleton, G., Howse, J., Chapman, P., Delaney, A., Burton, J., Oliver, I.: Formalizing concept diagrams. In: Visual Languages and Computing, pp. 182–187. Knowledge Systems Institute (2013)
39. Stockdill, A., Raggi, D., Jamnik, M., Garcia Garcia, G., Sutherland, H., Cheng, P., Sarkar, A.: Correspondence-based analogies for choosing problem representations. In: C. Anslow, F. Hermans, S. Tanimoto (eds.) IEEE Symposium on Visual Languages and Human-Centric Computing, VL/HCC 2020, pp. 1–5. IEEE (2020)
40. Urbas, M., Jamnik, M.: Diabelli: A heterogeneous proof system. In: B. Gramlich, D. Miller, U. Sattler (eds.) IJCAR, *Lecture Notes in Artificial Intelligence*, vol. 7364, pp. 559–566. Springer (2012)
41. Urbas, M., Jamnik, M.: A framework for heterogeneous reasoning in formal and informal domains. In: T. Dwyer, H. Purchase, A. Delaney (eds.) Diagrams, *Lecture Notes in Computer Science*, vol. 8578, pp. 277–292. Springer (2014)
42. Urbas, M., Jamnik, M., Stapleton, G.: Speedith: A reasoner for spider diagrams. Journal of Logic, Language and Information **24**(4), 487–540 (2015)
43. Reasoning in OWL 2 RL and RDF graphs using rules. https://www.w3.org/TR/owl2-profiles/#Reasoning_in_OWL_2_RL_and_RDF_Graphs_using_Rules. Retrieved Dec 2019

Chapter 8
From Mathematical Reasoning to Crises in Different Languages: The Application of Failure-Driven Reasoning to Ontologies and Data

Fiona McNeill

Abstract Reasoning about failure has been a central pillar of DReaM Group research for a long time. But failure happens not just in maths but in all kinds of spheres. While failure can—and often does—occur during human communication, people are actually pretty good at identifying and correcting errors, and at communicating effectively even when they do not completely understand one another or have a different world view. But the ability to facilitate automated communication—for example, in peer-to-peer systems, or through automated data identification and integration—is difficult because misalignment and heterogeneity are common. In this chapter, I discuss my work over the years within the DReaM group, looking at different aspects of this problem.

8.1 Introduction

I first drifted into the DReaM group during my masters, when I was one of six students (with fellow future DReaMers Alison Pease and Dan Winterstein) taking Alan Bundy's Advanced Automated Reasoning course. Although at the time we were terrified by the grown up things he made us do (lead classes, give our own opinions on published work that had been done by real researchers, etc.), it was an excellent introduction to research life. I did my MSc dissertation with Alan and Jacques Fleuriot looking at dependencies in theorem proving to help determine the impact of making changes, from which I chiefly remember how long I spent developing perfect colours to highlight different kinds of dependencies. I then swanned off to Fiji and Australasia for a year off, leaving Alan to find me some funding to do a PhD. I got his email confirming his success in this while drinking cocktails in a beach bar in Ko Chang, Thailand, and then made my leisurely way

F. McNeill (✉)
University of Edinburgh, Edinburgh, UK
e-mail: f.j.mcneill@ed.ac.uk

G. Michaelson (eds.), *Mathematical Reasoning: The History and Impact of the DReaM Group*, https://doi.org/10.1007/978-3-030-77879-8_8

back overland through Asia and Europe to take up my PhD position in September 2001, settling into my South Bridge office about a year before it burnt down.

Although my inclination to travel never went away (I once got to a conference in Vancouver by taking the train from Edinburgh to Vladivostok, sailing to Japan, and then flying on from Tokyo), once ensconced in the DReaM group, I determined to remain and—though moving between different DReaM group sites, taking up a lectureship at Heriot-Watt before eventually moving back to the University of Edinburgh for a Readership—I always have.

Two key focuses that have underpinned a lot of what has been achieved within the DReaM group are:

- The application of failure-driven reasoning, in which reasoning about the circumstances around some kind of failure can be used to analyse potential problems in representation or reasoning
- The tension between representation and reasoning, where the richer and deeper the representation used to express information, the more difficult it is to reason over that information.

My research over the past 20 years has been rooted in these ideas and has focused on taking them outside the sphere of mathematical reasoning and into the world of language and communication.

In this chapter, I discuss the various different ways in which I have attempted to do this—each building on the last—and the domains to which this has been applied. I discuss the power—and also the limitations—of applying mathematical reasoning techniques in the complex and messy world of human language. In Sect. 8.2, I discuss how these approaches can help planning agents detect and correct ontology mismatches that are causing plan execution failure. In Sect. 8.3, I explain how some of the techniques developed in that work were integrated into a system to allow agents in a peer-to-peer network to map their abilities and goals onto interaction protocols. In Sect. 8.4, I take the structural and semantic matching techniques involved in that project and apply them to the automatic and semi-automatic rewriting of failed queries, and discuss the application of this in a crisis response and management environment. In Sect. 8.5, I consider the application of such techniques within and between specific domains, in which language is used in a domain-specific way, and between different natural languages. Finally, in Sect. 8.6, I consider where I will be applying this approach in future, and conclude this chapter.

8.2 Agents Reasoning About Their Ontologies

> "To learn, a learner needs to formulate plans, monitor the plan execution to detect violated expectations, and then diagnose and rectify errors which the disconfirming data reveal".
> Frederick Hayes-Roth (Hayes-Roth, 1983)

The first instantiation of these ideas was in my PhD. The central idea—initially from Alan and then developed between the two of us and my second supervisors, first Chris Walton and then Marco Schorlemmer—was that *representation is a fluent*. Just as in mathematical reasoning, the way in which representation is used has a profound effect on what can be said and proved, so too in the language-based world of ontologies. By considering representation to be fluid, we are interested in changes and mismatches not just in the facts themselves but in the language in which the facts are expressed and in the range of things that can be expressed in a given formalisation. There has been a significant amount of work done in the field of ontology matching (see, for example, [1]), but this mostly focuses on the semantics of terms as defined through is—a hierarchies, rather than looking at the language of the underlying representation. We were interested in the pragmatic question of how an entity using an ontology (i.e., a virtual agent) can use that ontology to successfully navigate its world and achieve its goals even though its ontology is unlikely to be a perfect representation of that environment. Because our goals are pragmatic, action is only instigated by failure—in our case, failure of plans to execute correctly that appeared to be workable—and repair is only considered necessary to the extent that it allows a replay of the plan execution so that that particular failure no longer occurs. We particularly focused on the domain of peer-to-peer agents, where plans are based on communication with other agents, and failure happens either when that communication breaks down or when, having assumed that the communication is successful, the expected outcome of the communication does not transpire.

The key differences between our approach and standard ontology matching are that:

- Ontology matching focuses on matching two different ontologies together, whereas we were interested only in one ontology—that of our agent—and how well it helps that agent to navigate its world. Feedback in our context came from interaction with other agents in a peer-to-peer system, but the same philosophy could apply to robots learning from the physical world, for example. In our world, an ontology is good enough if it allows the agent to achieve its goals within its domain.
- Most often, ontology matching creates maps between ontologies, leaving the ontologies themselves unchanged. We were interested in diagnosing and then implementing necessary changes to overcome obstacles, based on feedback around failure.
- We were interested in minimal matching and repair, concerned only with mismatches that are responsible for some kind of failure and anything else in the ontology that is directly affected by the repaired mismatch. As well as making the matching more tractable, this also has the advantage of requiring that only minimal parts of an ontology are shared, thus allowing for significant privacy.
- Since we were concerned with agent communication, we needed to think about semantic matching such as is commonly done in ontology matching—for example, when one agent uses the term *person* and the other uses the term *man*

are they talking about the same thing[1]? But our main focus was on mismatches in the representation language in which the ontology is written, which is not addressed in standard ontology matching.

Here, ORS—the Ontology Repair System—that we developed to address this problem is briefly described.

We are concerned with the resolution of the problem of ontological mismatch within a planning context, such as would be necessary for an agent to orchestrate Semantic Web services to reach a goal. An agent forms plans to achieve a goal based on its understanding of the domain and then attempts to execute these plans through communication with other agents. Planning in complex and dynamic environments is very difficult because any incomplete, incorrect, or out-of-date information can cause an inexecutable plan to be developed because the environment is changing while planning is being performed. However, by adopting our approach to ontological repair, these cases of plan execution failure can be considered to be opportunities to learn more about the domain through repairing a mismatched ontology. Executing plans, in our environment, is done by interacting with agents who can perform the necessary tasks of the plan: for example, buying a ticket is performed by successfully interacting with a ticket-selling agent. Information about the cause of failure is extracted from observation of the communication surrounding plan failure, augmented by further communication with the other agents involved. Once the point of failure has been located, repair techniques are implemented to fix the problem, and a new plan is developed using the updated ontology. This plan is, in general, more likely to succeed than the previous plan. This procedure is repeated until the goal is successfully reached, or until it becomes impossible to form a plan to achieve the goal from the updated ontology. Because the point is for an agent to update its ontology to better match the specific ontology of the agent it is currently interacting with, this does not represent a move towards a better ontology in any kind of global sense but rather an ability to interact correctly in the specific situation it is currently in. Further interaction with other agents may lead to repairs being undone or different changes made. Repair is used retrospectively rather than proactively. That is, ORS does not assume that unexpected communications are problematic and immediately investigate; instead, it stores unexpected communications—which we term *surprising questions*—and carries on regardless, only revisiting these surprising questions as part of the diagnostic process if failure occurs at some subsequent point. A surprising question is anything that does not completely align with a communication the agent expected to receive at a given point.

We consider that there are three essential elements to creating such a dynamic ontology repair system:

---

[1] As an aside, the brilliant *Invisible Women* by Caroline Criado Perez [2] is all about how we almost always do make this assumption—hence women being almost invisible in the data, with all kinds of negative consequences.

- The ability to link the relevant information about the underlying ontology to the plan
- The ability to use this information to diagnose the exact source of the problem where possible, or at least to narrow it down
- The ability to select and apply appropriate techniques for altering the ontology

We focused on agents that were using first-order ontology representations. This is more expressive than most ontological representations, but a richer representation is necessary for agents forming complex plans. The tension between a rich representation—necessary for representing complex relationship between objects and individuals—and a representation that can be reasoned over efficiently—necessary for complex planning—is discussed in [3]. Our solution was to use KIF, a first-order ontology language, for representing agents' knowledge and then translate into PDDL, a simplified representation, for developing plans. We developed the notion of pseudo-variables to create a closed-world representation of an open-world environment. Pseudo-variables are objects that are presented to the planner to be constants, thus making plan development tractable, but which were interpreted within the ontology as variables.

We defined four kinds of mismatches—both abstractions (where we need to be more general) and refinements (where we need to be more specific), based on the abstractions defined in [4]:

1. **Predicate abstraction** (conversely, predicate refinement), where predicate names are matched in some uniform way: e.g., $(Bottle?X), (Cup?X)$ map onto $(Container?X)$
2. **Domain abstraction** (conversely, domain refinement), where constants and function symbols are matched in some uniform way: e.g., (Prime 3), (Prime 5) map onto (Prime Oddnumber)
3. **Propositional abstraction** (conversely, propositional refinement), where some or all of the arguments to predicates are dropped: e.g., (Abelian GroupA), (Abelian GroupB) map onto (Abelian)
4. **Precondition abstraction** (conversely, precondition refinement), where preconditions are dropped, e.g.: (Has Ticket Me) $\rightarrow$ (Can-Travel Me) maps onto (Can-Travel Me)

All potential changes to first-order terms or to planning rules formed of first-order pre- and postconditions can be matched by combining these repairs. In [5], we demonstrate that ORS will always diagnose a specific mismatch to be repaired. In most cases, this is a precise and correct diagnosis, though in some cases it is a guess at the most likely cause. ORS then attempts to implement a repair of the mismatch. In the case of abstraction—i.e., when detail needs to be removed—this can be done automatically. Automatic repair is also sometimes possible for refinement—when detail needs to be added—because full information of the missing details can be gleaned from agent communication. However, it is often not possible to make such repair automatically. For example, it is not difficult to diagnose precondition refinement—essentially, that a precondition is missing—

but it is usually impossible to determine automatically what this. Likewise, in the case of predicate refinement—where an argument needs to be added—it is possible to add an argument of the correct type automatically and instantiate it correctly for the specific instantiation where the mismatch was diagnosed, but if this predicate appears at other places in the agent's ontology, then it is not usually possible to determine automatically how to instantiate all the other instances.

Evaluating ORS across seven different ontologies containing 325 different mismatches, it correctly diagnosed 70.8% of the relevant mismatches—that is, mismatches that had some relevance to an automated system (so ignoring changing in commenting, for example) and mismatches that could be diagnosed in a planning environment (e.g., excluding the removal of instances from an ontology). The additional 29.2% of the mismatches highlighted extensions that could be made to improve the functionality of ORS.

8.3 Into Ontology Matching: The Open Knowledge Project

After my PhD, I spent a few months working with Alan, Harry Halpin, and Ewen Klein on a short project looking at comparing different versions of classic children's stories by extracting first-order terms from natural language and matching them [6] (this proved only to be feasible for very simple narratives) and then started work on the EU-funded Open Knowledge Project. The project was coordinated by Dave Robertson (a former DReaMer) and involved a wide range of brilliant people, some of whom became long-term collaborators and friends: Fausto Giunchiglia in Trento, Frank van Harmelen in Amsterdam (both former DReaMers), Carles Sierra and Marco Schorlemmer in Barcelona, Enrico Motta at the Open University and, as well as Dave, Paolo Besana and Nadine Osman in Edinburgh. Southampton, led by Nigel Shadbolt, also brought Tim Berners-Lee into the project—a fun colleague for my first job. Unfortunately, he was too busy to attend a single meeting of the project. This did not go down brilliantly—in the words of our EU contact, if you advertise a concert with Pavarotti, you had better produce him at some point or people will go away disappointed ...

Alan helpfully encouraged Dave to give me a job by suggesting Dave to take me along to Trento for a planning meeting when the proposal was being developed, and with both my former PhD second supervisors on the project and the central involvement of Fausto, on whose work much of my PhD was based, it was an obvious next step for me. Although I was based in Edinburgh throughout the project, I worked almost exclusively with Trento—primarily with Fausto and his group on matching, and also with Maurizio Marchese on Emergency Response, one of the two application domains.

The Open Knowledge project focused on facilitating interaction between peers in a Semantic Web environment. This was done by providing interaction models that could be shared and reused, and detailed how interactions with certain outcomes would unfold. Peers could choose to take on roles in the interactions if their skills/knowledge/abilities were aligned with the preconditions of the actions that role had to take, and the outcomes of the role were aligned with their goals. Preconditions and outcomes were expressed in first-order logic, so in order to determine the suitability of a role, peers needed a way to measure the closeness of first-order terms: their abilities and goals mapped to those of the role. The challenge for the matching part of the project was to develop a system that would take in first-order terms and output a map between those terms and a numerical judgement of how similar they were.

For this, we developed the Structure-Preserving Semantic Matching (SPSM) algorithm [7]. This was built on top of the S-Match system [8], an ontology matching system developed by Fausto's group in Trento. First-order terms were represented as trees, and the matching problem was broken into two steps: *node* matching, performed by S-Match, and *tree* matching, which takes the results of node matching as input and then uses a tree-edit distance (TED) algorithm to measure the distance between the trees. This produces two outputs: (i) a map between the two terms, so that the agent can understand how to interpret the requirements and outcomes of the role in terms of its abilities and goals; (ii) a score in [0 1] indicating the similarity. This allows an agent to judge whether its map to the role is *good enough* for it to want to go ahead with it. It also allows an agent to communicate to other agents how well it will be able to perform a role, and they can then decide whether they view it as a suitable collaborator for the given interaction model.

For example, consider an agent that is attempting to get information about wine—it wants to know about region, country, price, colour, and number of bottles. It finds an interaction model that involves exchanging information about wine, but it is ordered slightly differently. Figure 8.1 shows how this might be mapped.

SPSM calculates the cost of moving from one tree to another, considering the cost of the node mapping returned by S-Match and structural manipulation returned by the TED. Figure 8.2 shows the costs associated with this.

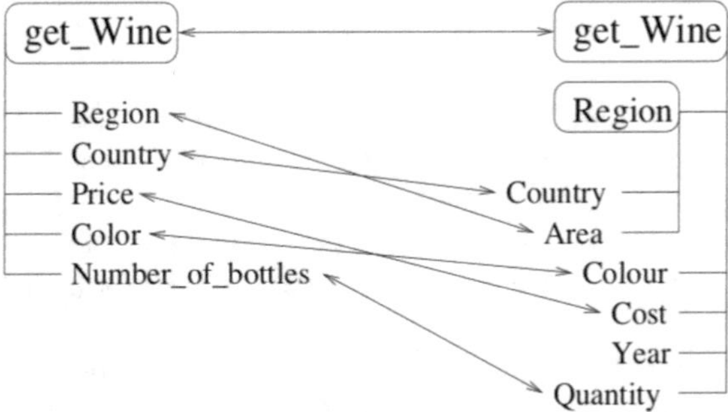

Fig. 8.1 Two approximately matched web services represented as trees: T1: get wine (Region, Country, Colour, Price, Number of bottles) and T2: get wine (Region (Country, Area), Colour, Cost, Year, Quantity). Functions are in rectangles with rounded corners; they are connected to their arguments by dashed lines. Node correspondences are indicated by arrows

Abstraction operations	Tree edit operations	Preconditions of operations	$Cost_{T1=T2}$	$Cost_{T1\sqsubseteq T2}$	$Cost_{T1\sqsupseteq T2}$
$t_1 \sqsupseteq_{Pd} t_2$	$a \rightarrow b$	$a \sqsupseteq b$; a and b correspond to predicates	1	∞	1
$t_1 \sqsupseteq_{D} t_2$	$a \rightarrow b$	$a \sqsupseteq b$; a and b correspond to functions or constants	1	∞	1
$t_1 \sqsupseteq_{P} t_2$	$a \rightarrow \lambda$	a corresponds to predicates, functions or constants	1	∞	1
$t_1 \sqsubseteq_{Pd} t_2$	$a \rightarrow b$	$a \sqsubseteq b$; a and b correspond to predicates	1	1	∞
$t_1 \sqsubseteq_{D} t_2$	$a \rightarrow b$	$a \sqsubseteq b$; a and b correspond to functions or constants	1	1	∞
$t_1 \sqsubseteq_{P} t_2$	$a \rightarrow \lambda$	a corresponds to predicates, functions or constants	1	1	∞
$t_1 = t_2$	$a = b$	$a = b$; a and b correspond to predicates, functions or constants	0	0	0

Fig. 8.2 The correspondence between abstraction operations, tree-edit operations, and costs

The similarity between trees is calculated by considering the cost of moving from one to the other, as calculated in Equation 8.1.

$$Cost = min \sum_{i \in S} k_i * Cost_i, \tag{8.1}$$

where S stands for the set of the allowed tree-edit operations; k_i stands for the number of i(th) operations necessary to convert one tree into the other, and $Cost_i$ defines the cost of the i(th) operation. Our goal here is to define the $Cost_i$ in a way that models the semantic distance.

We exploit the following equation to convert the distance produced by a tree-edit distance into the similarity score:

$$TreeSim = 1 - \frac{Cost}{max(T_1, T_2)}, \tag{8.2}$$

where $Cost$ is taken from Eq. 8.1 and is normalised by the size of the biggest tree. Note that for the special case of $Cost$ equal to ∞, TreeSim is estimated as 0. Finally, the highest value of TreeSim computed for $Cost_{T1=T2}$, $Cost_{T1\leq T2}$ and $Cost_{T1\geq T2}$ is selected as the one ultimately returned. For example, in the case of example of Fig. 8.1, when we match $T1$ with $T2$, this would be 0.62 for both $Cost_{T1=T2}$ and $Cost_{T1\leq T2}$.

Beyond the Open Knowledge project, the SPSM algorithm has been used in many contexts. A significant extension was to integrate it with evaluations of trust to provide algorithms for determining the best peer to interact with based on both their skills (evaluated by SPSM) and their reliability [9]. This integrated work was used, for example, to rank moderators in online sports sites, in supplier management, and in finding good sales managers.

8.4 Sharing Knowledge in Crises: Failure-Driven Query Rewriting

The fundamental idea of SPSM—determining similarity of structured semantic terms in a format that can be translated into trees—is fairly general, and over the next few years, I developed the concept in different domains. Alan and I had a few year-long grants after the Open Knowledge project, some with him as PI, some with me as PI, and some with him pretending to be PI for grants I wrote because of stringent EPSRC rules against PIs funding themselves from research grants. In 2013, I made the biggest move of my academic life—seven miles west to Heriot-Watt University in the Pentland hills. I took this research with me and developed it over the years into the basis of an approach to dynamic querying during crises.

Fast, effective data sharing is a requirement in many fields—for example, during crisis situations, in online retailing, and many Semantic Web applications. This will often take the form of queries from one organisation being sent to data sources belonging to other organisations. Automated query answering is a well-studied field. But successful querying of a data source depends on a good understanding of that data source, thereby ensuring that the schema and the data of the query correctly align with the schema and data of the queried data source. If data querying is part of an automated process, such knowledge depends on being able to anticipate in advance exactly what data sources will be relevant and knowing accurately what the schema and data representations of that source will be at the time of querying. If such knowledge is possible, then effective communication is best addressed by pre-alignment of data sources; in an ideal situation, these data sources would even use the same fixed vocabulary for easy integration. However, in the general case, such an approach is unrealistic. In a highly dynamic environment, it is usually not valid to assume we will know, in advance, exactly with whom we will need to interact or exactly what the context of this interaction will be. Assuming so precludes the possibility of dynamic interaction with new organisations, not anticipated at design time, and of interacting with known organisations that have updated or altered their data in some way. While pre-alignment is desirable where possible, depending only on this enormously limits the possible interactions during the response.

I addressed the problem through the development of the CHAIn (Combining Heterogeneous Agencies' Information) system [10], which built on the structural-semantic matching of the SPSM algorithm and applied it to query rewriting. It can be used by the owner of a data source to interactively formulate appropriate responses to incoming queries, even when these queries fail to match the data source at the schema level and/or the data level.[2]

CHAIn has the capability to perform in a fully automated manner and send responses ranked purely through automated matching. But it can also be used inter-

[2] A mismatch at the *schema level* is one where the structure of the data differs: for example, the columns of a database are in a different order. A mismatch at the *data level* is one where the schemas match at this particular point, but the specific data differs.

actively, with humans employed by the data owners filtering the automated ranking in order to return the best options. This is more expensive than full automation but has the advantage of providing context to matching that would otherwise be context-free. There is a lot of knowledge and intuition within institutions about how specific terminology is used locally and what aspects of data are crucial for specific tasks, and this local knowledge is hard or impossible to encode in a data source. Because of this innate unformalised knowledge, however high quality the automated matching is, better results are achieved with some input from humans who have this knowledge, and CHAIn then acts as a tool for fast, efficient human interaction with large data sources. The choice whether to run CHAIn fully automatically or interactively would depend on the task: whether speed or precision is preferred, whether there are suitably qualified humans available, etc.

In many (or most) cases, there will not be a single entry in the target data source that provides an exact match for the incoming query; it will often be the case that there are many approximate matches. Matches in CHAIn receive a numerical score based on the SPSM algorithm, and all matches that pass a given threshold are returned to the human user, ranked according to score. Effective ranking is essential; the task of the human is made tractable by the automated system returning a small number of highly relevant responses appropriately ranked, so that the human user can quickly hone in on the best response. If CHAIn is operating fully automatically, the highest ranked match is assumed to be the correct one.

The lifecycle of CHAIn is as follows:

0. The schemas of the target organisation D are extracted and translated into first-order terms, $D_{fol} = d_1, d_2, \ldots, d_n$, where n is the number of terms in the dataset. This will be done for every dataset the target organisation owns. This happens offline when CHAIn is installed locally. The rest of the process of CHAIn is triggered by an incoming query failing.
1. Once a failed query Q is sent to CHAIn, CHAIn extracts a first-order term, Q_{fol} from the query.
2. The predicate P of this first-order term is extracted and used to narrow down the datasets D of the target organisation to a small subset of data NDS, and their corresponding representations in first-order terms, $nds_1, nds_2, \ldots, nds_m, \forall nds_x \in D_{fol}, m \leq n$, that are potentially relevant.
3. Q_{fol} is sent pairwise with all $dx_{fol}, x = [1, \ldots, n]$ to the Structure-Preserving Semantic Matching algorithm (SPSM). SPSM determines mappings M_x between each pair and gives them a score $S_x \in [0\ 1]$. If $S_x > T$, where T is a given threshold $\in [0\ 1]$, SPSM returns a potential match PM_x, where $PM_x = \{nds_x, M_x, S_x\}$.
4. The returned matches are ordered according to their scores S_x. If desired, these results can be passed to a human in the target organisation, who may reorder the list and reject some or all of the proposed matches, to create a new list L. This results in a list L of ranked potential matches to Q_{fol}.
5. The potential matches in L are rewritten into queries based on the format of the original query Q and sent one by one to datasets. For each query Q_x that returns a

set of responses R_x, a list containing M_x and R_x is added to a list R of responses for the querying organisation.
6. Queries derived from L that fail to return a response are removed from L and returned to CHAIn for rewriting mismatches at the data level. If these exceed the threshold T, they are either added automatically to L or again sent to a human if required and added to L on the human user's approval.
7. The list L of results is returned to the querying organisation.

While the potential application of this is pretty general, my main interest in its application was within the crisis management domain. This is a domain particularly suited to this approach as, although extensive pre-event preparedness is essential, crisis events are characterised by their unpredictability, and layering a dynamic response to unexpected events, including engaging with unexpected partners, on top of pre-crisis planning is always necessary. It is extremely common for post-disaster reports to highlight a failure to share data effectively: crucial information that would have altered key decisions was out there but was not with the decision-makers at the crunch moment. This is because it is extremely difficult to source and consume any and all potentially relevant data, in part because this data is heterogeneous on multiple levels, making it hard to use automated methods to identify it. This is the problem I am attempting to solve.

I worked with many responders, particularly within the ISCRAM[3] community to develop this concept. My main focus has been on the technical underpinning of the matching and integrating of heterogeneous data sources, but the development of functional tools that could be relied on during a crisis is multi-faceted and depends on various technologies. For example, one big issue for responders is understanding where data comes from and how it was created so that they can make judgements about what data they want to trust. This led to a joint project with Newcastle and Coventry Universities during which we developed the CEM-DIT[4] system [11], which looked at integrating information about relevance of data (through the CHAIn system) with graphical depictions of the provenance of the data—for example, information such as which organisation provided the data, how figures used in the data were calculated, which tools were used to create images, and so on. Decision-makers in crises can explore the provenance at different depths until they are satisfied that they can make a judgement about the reliability of that data.

Another complicated aspect of this kind of data matching, and of matching in general, is evaluation. When matching is approximate and ranked, how is it possible to validate that the matches returned are the best ones available, and ranked in a way that is optimal? The main attempt to address this is the Ontology Alignment Evaluation Initiative (OAEI), which is run annually at the Ontology Matching Workshop. We make use of this resource as much as we can, but it has to be

[3]Information Systems for Crisis Response and Management.

[4]Communication in Emergency Management through Data Integration and Trust.

repurposed fairly significantly as it is not designed to evaluate the problem we are working with. We therefore developed a system that automatically corrupts queries according to various criteria in order to generate test sets for evaluation of CHAIn [12]. This is a useful tool but still limited in significant ways because to be effective it requires extensive lexicographic input to ensure the corruptions are plausible and valid. Again, we use WordNet for this, but the tool would be more effective where good-quality domain-dependent resources were available.

8.5 Multi-Lingual and Multi-Domain Matching

My work has primarily focused on English-language data and depends on general-purpose resources such as WordNet.[5] However, this is limited. One thing that became apparent during my work in crisis management is that language is used in a particular way in that domain—for example, words in general use in the language being used in a more specific, restricted way within the domain, and the use of jargon and terminology that are mainstream in the domain but unknown outwith. Automated matching tools with no way of interpreting and accounting for this domain specificity will produce results of limited functionality. Francisco Quesada-Real, a PhD student, looked into the application of this within the medical domain [13]. A different but related problem is that of integrating sources written in different languages. This has relevance in the crisis management world, where international and cross-border responses often involve organisations that hold resources in multiple languages. My work on multi-lingual and domain matching has been done jointly with Fausto Giunchiglia's KnowDive group in Trento, particularly with Gábor Bella [14]. It feeds into the work of the KnowDive group on the development of DataScientia—a soon-to-be-established not-for-profit organisation whose ultimate aim is the creation of a grass-roots community centred around the development and dissemination of a unitary knowledge-driven understanding of the people's diversity, as it is represented by the data in the Internet.[6]

Ontologies and other knowledge organisation systems, while usually serving a purpose of standardisation or generalisation, stem from local needs and practices. By local we understand within an administrative unit such as a country or a region as well as within an application domain such as medicine or transport. Accordingly, ontologies tend to target specific domains, and the labels annotating their elements—concepts, relations, metadata—tend to be expressed in the local language. This is especially true for lightweight ontologies [15]: classification hierarchies, taxonomies, and other tree-structured data schemas widely used around the world as simple, well-understood, semi-formal resources for knowledge organisation. Such resources often play normative roles on the national level in public

[5]https://wordnet.princeton.edu.

[6]http://datascientia.disi.unitn.it.

services, industry, or commerce, as a means for classification (of documents, books, open data, commercial products, web pages, etc.) as well as being sources of shared vocabularies for actors cooperating in a given domain.

Activities on supra-national levels such as international trade and mobility need to rely on the interoperability and integration of knowledge organisation resources across countries, languages, and sometimes across domains. Cross-lingual matching is a specific case of language-aware matching when ontologies in different languages need to be aligned. Likewise, cross-domain matching is used to match ontologies pertaining to different domains of knowledge. An example of a simultaneously cross-lingual and cross-domain matching problem is the case of cross-border emergency response where responders from different countries and from different domains (geography, geology, medicine, police, military, transportation, etc.) need to share data. In [16], we apply the domain-aware matching approach presented in this chapter to this particular use case.

State-of-the-art cross-lingual matchers invariably use translation-based techniques—most often online machine translation services from Microsoft or Google—in order to reduce the problem of multilingualism to the well-researched problem of monolingual English-to-English matching (e.g., [17–19]). With the constant improvement of such services, translation-based matchers are able to provide usable results and are able to deal with a wide range of languages. State-of-the-art machine translators today mainly use statistical methods and are trained on large amounts of bilingual parallel or comparable corpora for each language pair they support.

A known problem of statistical machine translation, however, is the decrease of translation accuracy on corpora significantly different from those on which the system was trained. This typically happens on domain classifications and ontologies that contain specialised terminology. The adaptation of a statistical system to a new domain requires re-training on corpora extended with a significant amount of domain-specific text (ideally bilingual parallel corpora that is hard to find). At the same time, the systems typically used by ontology matchers are online commercial services (such as Bing and Google Translate) that, while offering the best available translation quality, are not adaptable or customisable by the user.

The shortness of labels typically found in ontologies is another difficulty that state-of-the-art approaches face, as the sparseness of textual context within labels makes the translation task more error-prone. Furthermore, the often non-standard orthography and syntax of ontology labels—that we described in [9] as a form of specialised block language—make label parsing even harder.

We followed a different approach to language- and domain-aware matching, so far hardly investigated, that does not rely on external translation tools. The method is based on combining two types of resources: on the one hand, multi-lingual natural language processing tools that are adapted to the language of structured data, and, on the other hand, offline multi-lingual lexical databases connecting words and expressions of natural language to language-independent but domain-aware meanings.

Our work is motivated by the following considerations. First, while both approaches evoked above are resource-intensive, the types of resources they feed on are markedly different: on the one hand, machine translation requires large amounts of bilingual parallel or comparable corpora relevant to the target domain, and on the other hand, our approach uses lexical, terminological, and NLP resources for each supported language. In both cases, a wide range of open-source resources are already available. Based on their availability and conditions of use, for specific use cases, one approach or the other may be more cost-effective or faster to implement. Our knowledge-based label matching approach can thus be seen as an alternative when no good-quality language- or domain-specific machine translator is available. Second, we are interested in comparing the strengths and weaknesses of the two approaches, which turn out to be rather complementary. Our evaluations use two machine translation systems: Google Translate, currently the best available online translator, and Apertium, which is free and can also be used offline. We conduct evaluations on three language pairs: English–Spanish, English–Italian, and Spanish–Italian. Finally, based on the complementarity of the two approaches, we investigate the idea of combining them—using multi-lingual lexical resources on the one hand and machine translation on the other hand—into a single matcher. The resulting system, as demonstrated by our evaluation results, clearly outperforms either method alone.

The result of our work is implemented in NuSMATCH (NuSM for short), an upcoming release of the open-source SMATCH system with built-in capabilities for language- and domain-aware matching.

Gábor and I have also worked with Sabhal Mòr Ostaig at the University of the Highlands and Islands to develop a lexical resource for Gaelic, which is again feeding into the DataScientia project (publication in progress).

8.6 On to the Future

The challenges around data integration and matching are huge, and there are many different directions this work could be taken. One of my main interests at the moment is around how people interface with data. For example, if a decision-maker in a crisis sends out requests for specific data and gets multiple results back from a system like CHAIn, possibly with multiple annotations such as provenance data through the CEM-DIT system, what is the best way to facilitate that person effectively sorting through the data to be able to make optimal decisions? How can the right amount of data to provide be determined, so that users have access to what they need to make decisions, and the potential to explore that in more depth if they would like, without overwhelming them with more data than can be processed? What are the social barriers—for example, institutional disinclination to trust data from other organisations, and the need to navigate what happens when poor outcomes occur following decisions made with the support of such systems?

I am also interested in the application of these approaches to other domains, particularly education. Computing and STEM education is something that I, like Alan Bundy, have been involved in for much of my career, and there are interesting issues relating to data matching and integration that arise in the field. For example, many teachers develop high-quality teaching materials that could be of huge value to other teachers, but these are rarely shared widely, in part because this is difficult to do. To be widely accessible, such materials would need to be suitably marked up with relevant metadata around what they were about, what age range, ability level, etc., they were aimed at and so on. Sharing materials on a broad basis would require ontologies describing what kinds of things you might want to say about educational materials, and ontologies describing how different international educational systems and curricula relate to one another. Because there will be a huge variety in how teachers describe their material, sophisticated, domain-specific, interlingual matching would be necessary to make these materials fully accessible. This again raises interesting questions about how people interact with data: how does one engender trust and a willingness to engage with systems that automatically manipulate data?

The problems around the application of matching tools to databases, ontology, and natural language are profound—in part because they rely on humans labelling and describing their knowledge accurately, clearly, and consistently, which in practice hardly ever happens. But the potential to develop tools that can work with humans to help them navigate a world of massively heterogeneous data and support them in sharing data and knowledge more effectively is huge, and I look forward to working with DReaMers past and present towards solving these issues.

References

1. J. Euzenat and P. Shvaiko, *Ontology Matching*. Berlin, Heidelberg: Springer-Verlag, 2007.
2. C. C. Perez, *Invisible women: data bias in a world designed for men*. Vintage, 2019.
3. A. B. Fiona McNeill and C. Walton, "An automatic translator from KIF to PDDL," in *Proceedings of the Planning Special Interest Group (PlanSIG)*, 2004.
4. F. Giunchiglia and T. Walsh, "A theory of abstraction," *Artificial Intelligence*, vol. 56, 1992.
5. F. McNeill and A. Bundy, "Dynamic, automatic, first-order ontology repair by diagnosis of failed plan execution," *IJSWIS (International Journal on Semantic Web and Information Systems) special issue on Ontology Matching*, vol. 3, pp. 1–35, 2007.
6. F. McNeill, H. Halpin, E. Klein, and A. Bundy, "Merging stories with shallow semantics," in *Proceedings of the Knowledge Representation and Reasoning for Language Processing Workshop at the European Association for Computational Linguistics (EACL) conference (KRAQ 2006)*, 2006.
7. F. Giunchiglia, F. McNeill, M. Yatskevich, J. Pane, P. Besana, and P. Shvaiko, "Approximate structure-preserving semantic matching," in *On the Move to Meaningful Internet Systems: OTM 2008*, pp. 1217–1234, Springer, 2008.
8. F. Giunchiglia, P. Shvaiko, and M. Yatskevich, "S-Match: An algorithm and an implementation of semantic matching," in *ESWS*, vol. 3053, pp. 61–75, Springer, 2004.
9. N. Osman, C. Sierra, F. McNeill, J. Pane, and J. Debenham, "Trust and matching algorithms for selecting suitable agents," *ACM Trans. Intell. Syst. Technol.*, vol. 5, pp. 16:1–16:39, Jan. 2014.

10. F. McNeill and A. Gkanaitsou, "Dynamic data sharing from large data sources," in *Proceedings of the ISWC Workshop on Discovering Meaning on the go in Large Heterogenous Data*, (Boston, USA), November 2012.
11. F. McNeill, D. Bental, P. Missier, J. Steyn, T. Komar, and J. Bryans, "Communication in emergency management through data integration and trust: an introduction to the CEM-DIT system," in *In Proceedings of 16th International Conference on Information Systems for Crisis Response and Management (ISCRAM)*, 2018.
12. F. McNeill, D. Bental, A. J. Gray, S. Jedrzyejczyk, and A. Alsadeequi, "Generating corrupted data sources for the evaluation of matching systems," in *In Proceedings of the 14th International Workshop on Ontology Matching*, 2019.
13. F. J. Real, F. McNeill, G. Bella, and A. Bundy, "Improving dynamic information exchange in emergency response scenarios," in *In Proceedings of 15th International Conference on Information Systems for Crisis Response and Management (ISCRAM)*, 2017.
14. G. Bella, F. Giunchiglia, and F. McNeill, "Language and domain aware lightweight ontology matching," *Journal of Web Semantics*, vol. 43, pp. 1–17, 3 2017.
15. F. Giunchiglia, M. Marchese, and I. Zaihrayeu, "Encoding classifications into lightweight ontologies," in *Journal on Data Semantics VIII* (S. Spaccapietra, P. Atzeni, F. Fages, M.-S. Hacid, M. Kifer, J. Mylopoulos, B. Pernici, P. Shvaiko, J. Trujillo, and I. Zaihrayeu, eds.), (Berlin, Heidelberg), pp. 57–81, Springer Berlin Heidelberg, 2007.
16. G. Bella, A. Zamboni, and F. Giunchiglia, "Domain-based sense disambiguation in multilingual structured data," in *Proceedings of International Workshop on Diversity-Aware Artificial Intelligence (Diversity @ ECAI 2016)*, pp. 53–61, 8 2016.
17. G. Bella, F. Giunchiglia, A. AbuRa'ed, and F. McNeill, "A multilingual ontology matcher," in *Proceedings of the 10th Workshop on Ontology Matching*, CEUR Workshop Proceedings, pp. 13–24, CEUR-WS, 10 2015.
18. L. Bentivogli, A. Bocco, and E. Pianta, "ArchiWordNet: Integrating WordNet with domain-specific knowledge," 2004.
19. F. Bond and R. Foster, "Linking and extending an open multilingual WordNet," in *Proceedings of the 51st Annual Meeting of the Association for Computational Linguistics (Volume 1: Long Papers)*, (Sofia, Bulgaria), pp. 1352–1362, Association for Computational Linguistics, Aug. 2013.

Printed in the United States
by Baker & Taylor Publisher Services